BARRON'S

Outsmart the TOEFL®

TEST STRATEGIES AND TIPS

Pamela J. Sharpe, Ph.D.

Publishing Globally
NEW AGE INTERNATIONAL (P) LIMITED, PUBLISHERS
New Delhi • Bangalore • Chennai • Cochin • Guwahati
Hyderabad • Kolkata • Lucknow • Mumbai
Visit us at www.newagepublishers.com

First Indian Sub-Continent Edition: 2014
by New Age International (P) Ltd., Publishers

BARRON'S: OUTSMART® THE TOEFL TEST STRATEGIES AND TIPS, First Edition by Pamela J. Sharpe, Ph.D.

Originally Published in the *United States of America*.

BRANCHES

- **Bangalore** 37/10, 8th Cross (Near Hanuman Temple), Azad Nagar, Chamarajpet, Bangalore-560 018
  Tel.: (080) 26756823, Telefax: 26756820, E-mail: bangalore@newagepublishers.com
- **Chennai** 26, Damodaran Street, T. Nagar, Chennai-600 017
  Tel.: (044) 24353401, Telefax: 24351463, E-mail: chennai@newagepublishers.com
- **Cochin** CC-39/1016, Carrier Station Road, Ernakulam South, Cochin-682 016
  Tel.: (0484) 2377004, Telefax: 4051303, E-mail: cochin@newagepublishers.com
- **Guwahati** Hemsen Complex, Mohd. Shah Road, Paltan Bazar, Near Starline Hotel, Guwahati-781 008
  Tel.: (0361) 2513881, Telefax: 2543669, E-mail: guwahati@newagepublishers.com
- **Hyderabad** 105, 1st Floor, Madhiray Kaveri Tower, 3-2-19, Azam Jahi Road, Nimboliadda, Hyderabad-500 027
  Tel.: (040) 24652456, Telefax: 24652457, E-mail: hyderabad@newagepublishers.com
- **Kolkata** RDB Chambers (Formerly Lotus Cinema) 106A, 1st Floor, S.N. Banerjee Road, Kolkata-700 014
  Tel.: (033) 22273773, Telefax: 22275247, E-mail: kolkata@newagepublishers.com
- **Lucknow** 16-A, Jopling Road, Lucknow-226 001
  Tel.: (0522) 2209578, 4045297, Telefax: 2204098, E-mail: lucknow@newagepublishers.com
- **Mumbai** 142C, Victor House, Ground Floor, N.M. Joshi Marg, Lower Parel, Mumbai-400 013
  Tel.: (022) 24927869, Telefax: 24915415, E-mail: mumbai@newagepublishers.com
- **New Delhi** 22, Golden House, Daryaganj, New Delhi-110 002
  Tel.: (011) 23262368, 23262370, Telefax: 43551305, E-mail: sales@newagepublishers.com

---

ISBN : 978-81-224-3573-3

₹ 299.00

C-13-04-6881

Printed in India at Glorious Printers, Delhi.

PUBLISHING FOR ONE WORLD
NEW AGE INTERNATIONAL (P) LIMITED, PUBLISHERS
7/30 A, Daryaganj, New Delhi-110002
Visit us at **www.newagepublishers.com**

CONTENTS

3 STRATEGIES FOR THE LISTENING SECTION 63

4 STRATEGIES FOR THE SPEAKING SECTION 111

6 STRATEGIES FOR THE WRITING SECTION 217

7 AUDIO SCRIPTS AND ANSWER KEYS 295

INTERVIEW WITH DR. PAMELA SHARPE, TOEFL AUTHOR

Professor Lee:	Dr. Sharpe, I notice that you have taught at many universities in both the United States and abroad. To mention a few, I see the University of Florida, the Ohio State University, the University of Toledo, the University of Texas, and Northern Arizona University on your resume.
Dr. Sharpe:	Yes, I have been teaching for a long time.
Professor Lee:	How long have you been helping students prepare for the TOEFL?
Dr. Sharpe:	41 years.
Professor Lee:	And when was the TOEFL published?
Dr. Sharpe:	The TOEFL came out in 1963. So, if you do the math, you will see that I started my journey in TOEFL preparation only a few years after the test was published.
Professor Lee:	That's a long time. Do you ever get tired of the TOEFL?
Dr. Sharpe:	Not really. First of all, the test is revised constantly, so I am always challenged to keep up with the changes. And probably more importantly, my students are always new, and I enjoy helping them with their goals—the first of which is to score well on the TOEFL.
Professor Lee:	Let's talk about the test revision. What do you think about the iBT® TOEFL?
Dr. Sharpe:	I think it is the best TOEFL yet. It encourages students to prepare for real tasks that they will face in a university setting. Of course, no test is perfect, but this is so much better than the discrete-item testing that was the hallmark for so many years on versions of the TOEFL.
Professor Lee:	There is no Structure section on the iBT®. What is your opinion of that change?
Dr. Sharpe:	Well, that is a concern, isn't it? Theoretically, the grammar is tested within the context of the other sections. If you study the rubrics for evaluation, you will see that grammar is included in both the Speaking and the Writing sections. So, if grammar is part of the TOEFL preparation curriculum, then it could be a good plan. The problem is that the pendulum swings between direct instruction and contextual instruction for grammar, with logical consequences. I've seen this a number of times during my teaching career. And, given the kinds of shortcuts that are considered acceptable in texting and e-mailing, I think some direct instruction in grammar is still needed somewhere in the curriculum. Ultimately, the instruction should show results on the total iBT® TOEFL score.
Professor Lee:	What about the student who does well in class but consistently scores lower on the TOEFL than you would expect? Have you had students with this problem?
Dr. Sharpe:	Indeed I have. We all have. And it is particularly disappointing when that happens. I tend to weave motivation into both my classes and my books because it often takes longer than students expect for them to

score at the level that they require to pursue their goals. Now we know that some students insist on taking the TOEFL before they are ready. But you asked about students who are prepared but don't succeed. Sometimes these students panic in a test situation. For them, I find that the best solution is to help them have success on model tests and practice in a simulation of the test situation. Then I like to support them with positive thinking and relaxation techniques to use when they take the TOEFL. In addition to that, sometimes it is possible to locate a few problem areas on the TOEFL. Then we can work individually on those to help a student add the essential extra points.

Professor Lee: Let's talk about your books. I know that you have published more than a dozen, and quite a few of them are TOEFL preparation books. Do you write all of the books yourself or do you have a team of writers who help you?

Dr. Sharpe: I'm asked that a lot. And, although I have worked with writing teams on a number of consulting projects, I have always done all of the writing myself for my books.

Professor Lee: So you have written everything in all the editions of *Barron's TOEFL*? I believe there are 13 editions of that one.

Dr. Sharpe: Yes, I wrote all 13 editions, beginning with the first edition in 1975.

Professor Lee: What are you working on now?

Dr. Sharpe: I am always updating my books so that students will have the most recent information and formats. Right now I am working on new editions for *Barron's TOEFL iBT* and *Barron's Practice Exercises for the TOEFL* as well as *Barron's Pass Key to the TOEFL.* I am also working on a mobile phone App for the TOEFL.

Professor Lee: Say, that's exciting.

Dr. Sharpe: Yes, I am very excited about it. I think that the strategies will help my students, and the new format—an App for their phones—well, that should make it easier for them to study, and maybe a little more fun, too.

Professor Lee: Is there anything that you would like to share with other teachers who are involved in TOEFL preparation, apart from the "Teaching Tips" that we are publishing along with this interview?

Dr. Sharpe: Yes. I'd like to say to all of my colleagues who are teaching in a TOEFL curriculum, that we are doing important work. As long as the TOEFL retains its place as a requirement for university admission, scholarships, and professional licensing, then our students need us to guide them and support their goals. I think we have a very unique role to play in ESL and EFL throughout the world and I feel fortunate to have been a part of it for so long, I applaud the work of my fellow TOEFL teachers, and I wish them and their students continued success.

For more about Dr. Sharpe, Visit her website at *www.teflprep.com.*

TEACHING TIPS FOR TOEFL PREPARATION

Dr. Pamela J. Sharpe

1. Begin with a positive message.
2. Write three important goals for the class so that students can see them.
3. Arrange for practice tests to be taken in a lab or at home on the honor system.
4. Allow students to grade the Reading and Listening sections of their practice tests.
5. Ask students to write their questions on note cards and bring them to class.
6. When several students have the same question, prepare a short presentation.
7. Make overheads of test questions and show the students how you choose an answer. Think out loud to eliminate distracters.
8. Use class time to teach and practice academic skills.
9. Focus on Speaking and Writing sections in class.
10. Assign Reading and Listening sections as homework.
11. Don't worry about covering all the material in the book.
12. Provide counseling and encouragement as part of the class routine.

TEACHER OBSERVATIONS: THINKING OUT LOUD

The teacher has a large visual of a test question with four possible answers at the front where students can see it while they listen. The teacher begins to talk to herself as she looks at the question.

What kind of company was discussed in the case study?

Ⓐ A new technology venture
Ⓑ A long-established organization
Ⓒ A distribution business
Ⓓ A truck manufacturer

"The company is new, but Choice A is not correct because the lecturer said that the company delivers new technologies, but it is not a technology venture. Let's see. The company has been in business for 3 months, and that is not very long, so Choice B is not correct. Choice C looks like the correct answer, but I should read Choice D before I decide. It is true that the company owns a fleet of trucks, but it does not manufacture them. Yes. Choice C is the correct answer."

The teacher has effectively modeled how to make a good choice on the test.

Best wishes to you and to your TOEFL students!

ACKNOWLEDGMENTS

Every project is the result of collaboration, but this one was really a team effort. I am grateful to be part of such an outstanding team.

Bob O'Sullivan, Publisher, Test Prep Division, Barron's Educational Series, Inc.
Thank you for providing the opportunity to write the book and for both leading the team and allowing us each the creative freedom to do our best work.

Wayne Barr, Acquisitions Editor, Barron's Educational Series, Inc.
Thank you for taking the time to understand the purpose of the project and for supporting it.

Kristen Girardi, Editor, Barron's Educational Series, Inc.
Thank you for managing all of the moving parts with enthusiasm, patience, skill, and good cheer.

Debby Becak, Senior Production Manager, Barron's Educational Series, Inc.
Thank you for creating just the right designs to enhance and clarify the content for international readers.

Sara Black, Copy Editor
Thank you for identifying and correcting the inconsistencies before anyone else saw them.

Karen Watson, Graphic Designer, Keyboard Graphics
Thank you for designing the pages and making important adjustments to the format.

Kathy Telford, Proofreader, Proofreader's Plus
Thank you for locating all of those pesky errors that find their way into a manuscript.

Myrna Goldstein, Director, Are You in Your English File?®
Second Language Research Center
Thank you for field testing preliminary materials and providing superb suggestions.

John Rockwell, CEO, Rockwell Audio Media
Thank you and the talented studio voices for interpreting the audio scripts just the way that I heard them in my imagination.

Paul J. Orozco, Technology Consultant, iProducts
Thank you for opening my eyes to the possibilities for helping students with technologies.

Robert and Lillie Sharpe, Devoted Parents
Thank you for the lessons that live in my heart, even after your passing.

John Osterman, Award-Winning Husband
Thank you for being my forever partner in this project and everything else in my life.

1

ORIENTATION

What are TOEFL strategies?

Many books help you prepare for the TOEFL, but this book is different from all the others. *Outsmart the TOEFL: Test Strategies and Tips* helps you prepare for the TOEFL before you take the test and it helps you outsmart the TOEFL while you are actually taking the test. How is this possible? By using this book, you will learn the strategies that you need.

A strategy is a way to think, plan, and outsmart the opponent. When you take the TOEFL, it is you against the test. This book contains 100 strategies based on the most current TOEFL formats. The strategies are tips, tactics, and tricks to outsmart the TOEFL. The tips provide advice on what to do for certain situations that arise during testing. The tactics show you how to approach the most commonly tested questions. The tricks give you the advantage of knowing how raters who grade the TOEFL want you to answer.

What are the current TOEFL formats?

Internet-Based iBT® TOEFL

The TOEFL iBT tests your ability to understand and use English for academic purposes. There are four sections on the TOEFL, with special directions for each section.

Reading Section
The Reading section tests your ability to understand reading passages like those in college textbooks. There are three passages on the short format and four passages on the long format. After each passage, you will answer 12–14 questions about it.

Listening Section
The Listening section tests your ability to understand spoken English that is typical of interactions and academic speech on university campuses. During the test, you will listen to conversations, lectures, and discussions, and you will answer questions about them. There are two conversations, two lectures, and two discussions on the short format and three conversations, three lectures, and three discussions on the long format.

Speaking Section
The Speaking section tests your ability to communicate in English in an academic setting. During the test, you will be presented with six speaking questions. The questions ask for a response to a single question, a conversation, a talk, or a lecture.

Writing Section

The Writing section tests your ability to write essays in English similar to those that you would write in college courses. During the test, you will write one essay about an academic topic and one essay about a familiar topic.

Summary of TOEFL iBT Internet-Based Test

Reading	3–4 passages 700 words each	12–14 questions each	60–80 minutes
Listening	2–3 conversations 4–6 lectures	5 questions each 6 questions each	60–90 minutes
Break			10 minutes
Speaking	2 independent tasks 4 integrated tasks	1 question each 1 question each	20 minutes
Writing	1 integrated essay 1 independent essay	1 topic each 1 topic each	20 minutes 30 minutes

Paper-Based ITP® TOEFL

The TOEFL ITP (Institutional Testing Program) tests your ability to understand and use English for academic purposes. There are three sections on the ITP, with special directions for each section. An additional writing test is often administered after the three-section test.

Listening Comprehension Section

The Listening Comprehension section tests your ability to understand spoken English that is typical of conversations and talks that you might hear on a university campus. During the test, you will listen to dialogues, conversations, and talks and you will answer questions about them. There are usually thirty dialogues, two or three long conversations, and three or four talks on the ITP format.

Structure and Written Expression Section

The Structure and Written Expression section tests your ability to recognize grammar and usage in standard, written English. During the test, you will identify 15 correct sentences in the Structure part and you will find errors in 25 sentences in the Written Expression part.

Reading Comprehension Section

The Reading Comprehension section tests your ability to understand short reading passages on general topics. There are usually five passages in the section. After each passage, you will answer 8–12 questions about it.

TWE—Test of Written English
The TWE tests your ability to write an essay in English on a general topic. During the test, you will write one essay.

Summary of TOEFL ITP Paper-Based Test

Listening Comprehension	30 Dialogues 2–3 Long Conversations 3–4 Talks	1 question each 3–5 questions each 3–5 questions each	35 minutes
Structure and Written Expression	Structure Written Expression	15 questions 25 questions	25 minutes
Reading Comprehension	5 passages 300–350 words	8–12 questions each	55 minutes
Test of Written English	1 independent essay	1 topic	30 minutes

How should I plan my study time?

- **Start early.** Give yourself at least a month to study the strategies in this book.
- **Study 2–4 strategies a day.** If you study more strategies in one session, you may not retain them.
- **Go slowly.** Take your time. Think about each strategy as you complete the practice.

How can I use this book effectively?

- **Design.** Most of the time, the layout of this book is designed so that you see explanations and examples on the left side and practice pages on the right side. This allows you to refer to the explanations and examples while you are working on the practice activities. Take advantage of this design feature while you are studying.
- **Order.** The order of the strategies is important because you will be using previously learned strategies to acquire new strategies. It is best to work from the beginning through the end of the book.
- **Chapters.** If you are taking the iBT, you will not have a Structure section. You may skip Chapter 5. If you are taking the ITP, you will not have a Speaking section. You may skip Chapter 4.
- **Bonuses.** Some of the strategies will help you on more than one section of the TOEFL. Bonus strategies will be marked and referenced with stars at the end of the practice activity that corresponds to them. In general, bonus strategies are more important than the other strategies. Give special attention to the bonus strategies with the most stars.

How can I use the images in this book?

This is a QR Code, short for Quick Response Code. It is a two-dimensional bar code that links you to information. Think of it as a shortcut to get to another place. To use the QR Codes in this book, you need a smartphone such as an iPhone, Android, Blackberry, Windows and Symbian Nokia phone, or a similar smart mobile device. Tablets with cameras will also accept QR Codes.

QR Codes are becoming more popular throughout the world. If you don't have an application on your mobile devices, you will need to download one in order to use the QR Codes in this book. Most of them are free! If you don't have an App yet, check your favorite App store. Several free options are available.

First, hold the camera phone in front of the QR code until you can see it in your viewer. Next, scan the image in the book with your smartphone's camera. When you do this, you will take a shortcut to the information in the code. For example, you might see a web link on your phone's display or you might be taken directly to a special site. Try it on the QR Code above. See what happens!

If you don't have a smartphone, you can still access the information. Alternative ways to locate the supplementary information and sites will be provided beside each QR Code. You can still get there, but you won't have the shortcut!

Why are some passages in the strategies shorter than those on the official TOEFL?

The purpose of each passage is to focus on one strategy for the TOEFL. Sometimes this strategy occurs only once in a passage, either at the beginning, the middle, or the end of the passage. To use your time efficiently and focus your attention on the point in the passage where the strategy is used, some of the reading and listening passages are edited to a shorter length.

What other resources will help me to succeed on the TOEFL?

After you have completed the practice activities in *Barron's Outsmart the TOEFL*, it is a good idea to continue your TOEFL preparation by taking model tests that simulate the official TOEFL exam. The reading passages and the listening passages on the model tests are the same length as those on the official TOEFL exam. The following books contain full-length model tests:

Barron's TOEFL iBT 13th Edition with CD-ROM
7 model tests iBT format in the book
8 model tests iBT format on the computer CD-ROM

Barron's Pass Key to the TOEFL iBT 7th Edition
3 model tests iBT format in the book
2 model tests iBT format on the Audio CDs

Practice Exercises for the TOEFL 7th Edition
1 model test iBT format in the book and on the Audio CDs
1 model test PBT format in the book and on the Audio CDs

Barron's TOEFL iBT Superpack
1 *Barron's TOEFL iBT, 13th Edition* book
1 CD-ROM
10 Audio Compact DIscs
1 *Barron's Essential Words for the TOEFL, 4th Edition* book

When you register to take the official TOEFL, you are entitled to a free model test, which you can find on the registration website at *www.ets.org/toefl*. This should be the last model test that you take before your official TOEFL.

Remember to use the strategies that you have learned in this book when you take the model tests, and again, when you take the official TOEFL exam.

What if I have a question about the strategies in this book?

The author will respond by e-mail to questions about the strategies. Please refer to the number of the strategy in the subject line of your e-mail when you ask your question. For example:

```
To: sharpe@teflprep.com
Cc:
Subject: Speaking Strategy 4
From: ana@me.net
________________________________________

Write your question here.
```

TIPS FROM SUCCESSFUL TOEFL STUDENTS

WHAT TO DO BEFORE THE TOEFL

- Ask your English teacher if you are ready to take the TOEFL. Some people take it before they know enough English. The TOEFL is expensive, and it can be a bad experience to take the test too soon. Be patient with yourself. It takes a long time to learn a language.
- If you have taken the TOEFL before and you have a score report, then you know which section is most difficult for you. Spend more time practicing for that section.
- Find the room the day before your TOEFL. That way, when you are going to your test, you won't get lost and you will be less stressed out about getting there late.
- Don't take a lot of stuff with you to the test. You can't take anything into the test room with you, and you will probably have a very small locker to store your things.
- Take care of yourself. Get as much sleep as possible the night before the test, eat breakfast and take a small snack with you for the break, go to the bathroom just before you go into the test room, and wear layers of clothing so that you can take off a sweater or put it on to adjust to the temperature in the room.

2

STRATEGIES FOR THE READING SECTION

OVERVIEW OF THE iBT READING SECTION

The Reading section is Section 1 on the Internet-Based TOEFL. The Reading section tests your ability to understand reading passages like those in college textbooks. Each passage is about 700 words in length.

There are two formats for the Reading section. On the short format, you will read three passages. On the long format, you will read four passages. After each passage, you will answer 12–14 questions about it. Only three passages on the long format will be graded. The other passage is part of an experimental section for future tests. Because you will not know which passages will be graded, you must try to do your best on all of them. You may take notes while you read, but notes are not graded. You may use your notes to answer the questions. Some passages may include a word or phrase that is underlined in blue. Click on the word or phrase to see a glossary definition or explanation.

Choose the best answer for multiple-choice questions. Follow the directions on the screen for computer-assisted questions. Most questions are worth 1 point, but the last question in each passage is worth more than 1 point.

The Reading section is divided into passages. Click on **Next** to go to the next question. Click on **Back** to return to previous questions. You may return to previous questions for all of the passages.

You can click on **Review** to see a chart of the questions you have answered and the questions you have not answered in each passage. From this screen, you can return to the question you want to answer.

You will have 20 minutes to read each passage and answer the questions for that passage. You will have 60 minutes to complete all of the passages and answer all of the questions on the short format. You will have 80 minutes to complete all of the passages and answer all of the questions on the long format. A clock on the screen will show you how much time you have to complete the Reading section.

OVERVIEW OF THE ITP READING COMPREHENSION SECTION

The Reading Comprehension section is Section 3 on the Paper-Based ITP. The Reading Comprehension section tests your ability to understand general interest reading passages. The passages are about 300–350 words in length.

You may NOT take notes while you read and you may NOT write in your test book. You can refer to the passage while you are answering the questions.

Choose the best answer from four possible answers written in your test book. Then, on your answer sheet, fill in the oval that corresponds to the letter of the answer you have chosen.

You will have 55 minutes to read and answer all of the questions.

TIPS FROM SUCCESSFUL TOEFL STUDENTS

WHAT TO DO ON THE READING SECTION

- Don't get scared if you don't know much about the topic. All of the information you need to answer the questions is in the passage, and technical vocabulary is in the glossary.
- Read the first paragraph and last paragraph of the passage first to get the main idea. Then start the questions.
- Don't spend too much time on a very difficult question. It is possible to go back and change your answer if you have time at the end of the passage.
- If you run completely out of time at the end, answer the remaining questions with the same letter. That way, you can finish quickly, you don't have to think, and you have a chance to answer some of them correctly to add points to your score.
- When you finish the Reading section, stop thinking about it, even if you didn't do very well. Get ready to do your best on the next section.

Reading Strategy 1

Reorganize the computer area for reading

When you sit down at your assigned computer area, you will find a computer screen, a keyboard, a mouse, earphones, note paper, and pencils. For the Reading section, you will not need the earphones, but you will be asked to put them on to test them. After that, if they are comfortable, you can just leave them on. They may be useful because they block out a certain amount of noise. You will not need the keyboard, the paper, or the pencils until you begin the Listening section. Move them to the side of your desk, out of the way. Adjust the screen and mouse so that you can use them efficiently to read the passages and click on the correct answers.

EXAMPLE

1. Put on the headphones.
2. Move the keyboard, paper, and pencils to the side.
3. Adjust the screen and the mouse.

Practice 1

Clear your desk except for the following items: computer screen, keyboard, mouse, earphones, paper, and pencils. Put everything in the middle in front of the screen. Now, as you reorganize your materials, say something positive in your mind. You cannot listen to two things at the same time, so it is important to listen to positive thoughts instead of negative thoughts.

1. Put on the headphones. "I am ready to listen. I am ready to speak."
2. Move the keyboard, paper, and pencils to the side. "I am ready to take notes. I am ready to write."
3. Adjust the screen and the mouse. "I am ready to read."

Do this every time you practice for the TOEFL. Do this when you take the official test. You will clear your desk and your mind at the same time.

Now, find five things that are wrong with this photo. What do you need to change in order to be well organized and ready to begin the Reading section of the TOEFL? What will you say to yourself? Check your answers with those in the Answer Key on page 295.

★★**Bonus:** iBT Listening, Speaking
You will not need the keyboard until you begin the Writing section of the iBT. Once you have organized your computer area, you can leave it until you begin to write your essays.

Reading Strategy 2

Practice reading either on the computer or in print

Depending on the version of the test you will be taking, you will need to practice reading either on a computer screen or on a printed page. Many readers actually read slower on a computer screen than on paper. Even if you can read at the same speed on both, the process is slightly different. You should have experience reading on screen if you are preparing for the iBT. On the iBT, you will see the passage on the right side of the screen with the questions on the left. If you are preparing for the ITP, then you should practice reading on paper. On the ITP, you will see the passage across the page with the questions below it.

EXAMPLE PASSAGE iBT

Dr. Jekyll and Mr. Hyde

The Strange Case of Dr. Jekyll and Mr. Hyde by Robert Louis Stevenson is the story of a London lawyer, Mr. Utterson, who investigates the relationship between his good friend, Dr. Henry Jekyll, and a mysterious acquaintance, Mr. Hyde. In fact, both Jekyll and Hyde are the same person with two distinct personalities. Dr. Jekyll is a respectable member of society who becomes attracted to a sinister lifestyle and invents a potion that transforms him at will into a monster, the savage Mr. Hyde. Eventually, the doctor loses his ability to control the transformations and finds that the moral side of his personality, Dr. Jekyll, is becoming weaker and weaker under the influence of the sinister Mr. Hyde.

Although much attention has been given to the Jekyll and Hyde characters, and their clear symbolic references to the good and evil in every man, the lawyer is also a character that deserves study. Mr. Utterson is the narrator and the character in whom all of the other characters confide. It can be argued that he also represents the epitome of the Victorian English gentleman. It is his nature to preserve social order even if it means that he must ignore the uncivilized activities to which he becomes a witness. He prefers to protect his friend's reputation by suppressing the truth about him because loyalty and the avoidance of scandal are of primary importance in the social code of his class. The importance of preserving appearances, even when it requires hiding indecent behavior is a major theme of the novel and one that is personified by the lawyer. Faced with the option of engaging in gossip and accusing his friend or ignoring the perverse activities that he has become aware of, Mr. Utterson decides to deny the obvious, even when he directly observes one of the physical transformations of his friend's appearance.

Stevenson's novel was written and read during the late 1800s, a time when Victorian society required a strict adherence to social customs as established and perpetuated by the upper class. It was perhaps this rigid morality that encouraged Dr. Jekyll's curiosity about unacceptable behavior, and Mr. Utterson's silence when he saw it.

EXAMPLE PASSAGE ITP

Dr. Jekyll and Mr. Hyde

The Strange Case of Dr. Jekyll and Mr. Hyde by Robert Louis Stevenson is the story of a London lawyer, Mr. Utterson, who investigates the relationship between his good friend, Dr. Henry Jekyll, and a mysterious acquaintance, Mr. Hyde. In fact, both Jekyll and Hyde are the same person with two distinct personalities. Dr. Jekyll is a respectable member of society who becomes attracted to a sinister lifestyle and invents a potion that transforms him at will into a monster, the savage Mr. Hyde. Eventually, the doctor loses his ability to control the transformations and finds that the moral side of his personality, Dr. Jekyll, is becoming weaker and weaker under the influence of the sinister Mr. Hyde.

Although much attention has been given to the Jekyll and Hyde characters, and their clear symbolic references to the good and evil in every man, the lawyer is also a character that deserves study. Mr. Utterson is the narrator and the character in whom all of the other characters confide. It can be argued that he also represents the epitome of the Victorian English gentleman. It is his nature to preserve social order even if it means that he must ignore the uncivilized activities to which he becomes a witness. He prefers to protect his friend's reputation by suppressing the truth about him because loyalty and the avoidance of scandal are of primary importance in the social code of his class. The importance of preserving appearances, even when it requires hiding indecent behavior is a major theme of the novel and one that is personified by the lawyer. Faced with the option of engaging in gossip and accusing his friend or ignoring the perverse activities that he has become aware of, Mr. Utterson decides to deny the obvious, even when he directly observes one of the physical transformations of his friend's appearance.

Stevenson's novel was written and read during the late 1800s, a time when Victorian society required a strict adherence to social customs as established and perpetuated by the upper class. It was perhaps this rigid morality that encouraged Dr. Jekyll's curiosity about unacceptable behavior, and Mr. Utterson's silence when he saw it.

Practice 2

Encyclopedias are especially helpful for reading practice because they contain extensive academic content and the reading difficulty level is about right. If you buy one that is several years old, it should be reasonably priced. Most of the information is the same as that of a new version, and since you are using it for reading practice, not for research, the older content is not a problem.

If you are practicing for the iBT, you can find free encyclopedias on the Internet or you can buy an encyclopedia on a CD-ROM so that you will be able to read from a screen. If you are practicing for the ITP, you can find older encyclopedias at a used bookstore.

Make a list of three options that you plan to use for reading practice. Check for additional suggestions in the Answer Key on page 295.

1. ______________________________

2. ______________________________

3. ______________________________

★★**Bonus:** iBT Speaking, Writing
The iBT includes reading passages in both the Speaking and Writing sections. Reading practice on screen will improve your scores on three sections of the iBT.

Reading Strategy 3

Glance at the images for orientation

When you see an image in the Reading section, you should not spend much time studying it. Just glance at the image and continue immediately to the passage. The information in the image will be repeated in the text, and the questions will refer to paragraphs, not to images. Most of the passages do not include images, but when they do, use them for orientation, and move on quickly.

EXAMPLE IMAGE

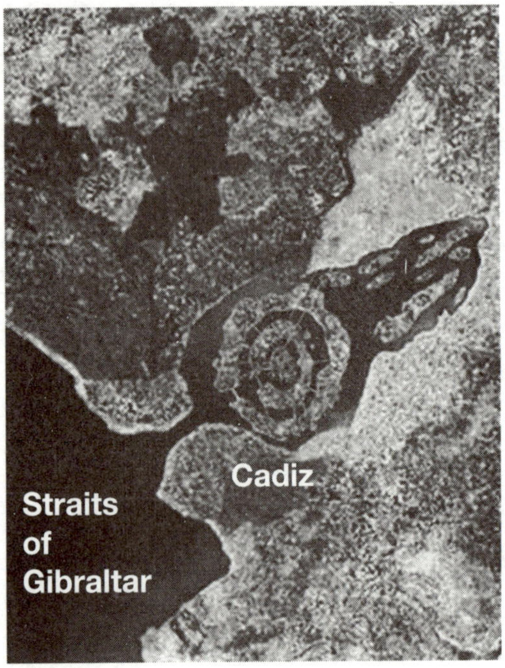

EXAMPLE PASSAGE

Atlantis

The lost city of Atlantis has been the topic of speculation and hypothesis since Plato first described it in his dialogues in 360 B.C.E. According to the legend, it was an island larger than Libya or Asia located near the Pillars of Hercules, which is now called the Straits of Gibraltar, off the southern coast of Spain. Presumably, the mythical city was totally destroyed by earthquakes and floods in a single night more than 9,000 years before Plato recorded the story.

Based on Plato's work, many historians and adventurers have proposed locations where Atlantis may have been submerged. Recently, technology has allowed them to see under the ocean in remarkable new ways. Previous claims for the discovery of the lost city have been made, identifying sites off the coast of Africa, in the Aegean Sea, the Mediterranean, and even in the mid Atlantic.

Using digital mapping and satellite imaging, an international team led by Dr. Richard Freund, professor and archaeologist from the University of Hartford, has uncovered what appears to be a city buried in a marsh in the south of Spain near Cadiz, about 60 miles from the coast. This is interesting in terms of Plato's record, since Atlantis was supposed to face the city of Gadera, and Cadiz is the modern name for this ancient city.

In addition to the man-made patterns of a metropolis, the team has located a large number of memorials nearby, constructed in the image of the buried city, which they believe the survivors may have built as remembrances of their lost homeland. If the team can identify geological formations that correspond to the descriptions by Plato of a ringed city, and if they can locate and date artifacts from the time period when Atlantis would have been a bustling civilization, then one of the great mysteries of the world will be solved.

Practice 3

Locate the text in the reading passage that identifies the locations on the map. Do not allow your mind to focus on the image at the beginning of the passage. Check your answers with those in the Answer Key on page 296.

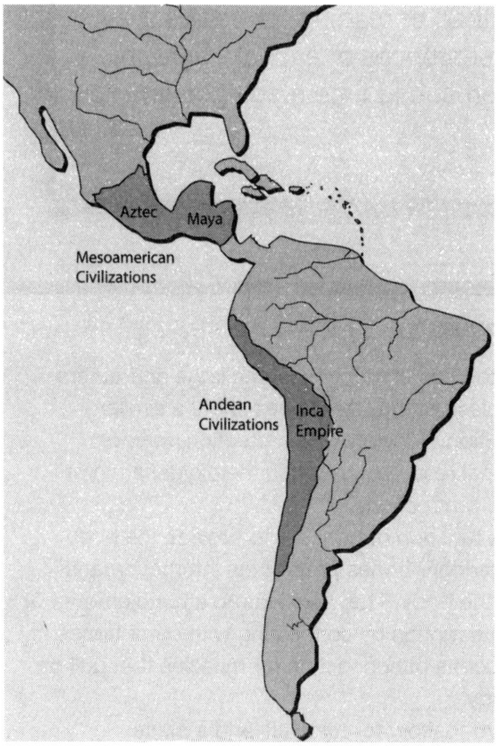

Pre-Columbian Civilizations

Although several other important cultures flourished, three major empires extended their influence over large regions in the Americas prior to the exploration and conquest by European powers in the sixteenth century. The Aztecs dominated the valley of Mexico where they intermarried with the Toltec nobility and systematically conquered smaller, weaker tribes from neighboring city states, consolidating their powerful kingdom under the rule of one chief. The subjugation of these regional city states allowed the Aztecs to exact tribute from more than half of the population of what is now Mexico, and assured their domination of Western Mesoamerica.

The Mayan culture stretched from El Salvador, Honduras, and Guatemala into central Mexico. During the classical period from 250 to 900 C.E., the construction of large urban areas identified their independent city states and dominated the landscape in Eastern Mesoamerica. Unlike the Aztec empire, there was no single Mayan political center. Because the Maya could not be overthrown by attacking a capital city where a centralized political system was in place, they remained strong and were able to survive invasions by competing tribes, exerting tremendous intellectual, if not occupational, influence throughout the region.

In what is now South America, the Incas established the largest empire of the Pre-Columbian cultures, uniting four regions in the Andes, including territories in what are now Ecuador, Colombia, Northern Chile and Argentina, Bolivia, and Peru. Ruling from a tribal base in Cuzco, the Inca emperor was considered the representative on Earth of one of the most revered gods, Inti, the sun god. The conquered tribes in the four regions maintained limited cultural identity under the rule of the Inca emperor but were, for the most part, integrated through peaceful negotiations and alliances with covenants of loyalty. Often the children of the ruling classes in the four regions were taken to Cuzco for their education, and daughters of the Inca nobility were married to powerful families in distant corners of the empire. If these peaceful methods proved inadequate, the formidable armies of the empire could easily subdue uncooperative tribes, execute the local leaders, and annex their cities.

Reading Strategy 4

Skim the passage for an overview

To *skim* means to look for general information in a passage. You will see each reading passage on the right side of the first screen without any questions on the left side. In order to progress to the questions, you must scroll through the entire passage. While you are scrolling, you can skim quickly for general information instead of reading for details. Pay special attention to titles, headings, and the first and last sentences of each paragraph. The passage will reappear with each question, and you can spend time reading each paragraph for details at that point.

EXAMPLE PASSAGE

The Function of Bones

The human skeleton has 206 bones, some large and others quite small, but regardless of the size, bones share a similar structure and composition. They provide protection, support movement, store mineral reserves, maintain the oxygenation of blood, and nurture cell production.

The most obvious function of bones is to provide the body with strength and protection. Bones protect the internal organs and provide shape to the body. They can absorb a huge amount of force, and they facilitate motion by connecting with other bones at joints and serving as points of connection for muscles that pull on bones to move the body.

Bones are also a repository for minerals and a place where they are constantly being renewed. Approximately every nine months, the mineral content of bone is depleted and replaced. Several minerals are of special importance. Calcium and phosphorus in particular are stored in bones and must be maintained in the blood at a stable level. What happens when the diet is depleted of these minerals is that they are removed from the bones to keep the necessary levels constant in the bloodstream. Moreover, when physical exercise is reduced, as in the case of injury or illness, then minerals are passed into the bloodstream from the bones, elevating the mineral content to levels that can cause kidney stones or other calcium deposits to occur in the organs.

Bones are also significant in the functioning of the circulatory system. Red blood cells, called erythrocytes, grow and mature in the bone marrow where they attract hemoglobin, which serves to carry oxygen in the blood. Because they live for only about four months, bone marrow must constantly reproduce erythrocytes to maintain healthy blood oxygen levels.

Finally, bone marrow serves as the site where lymphocyte precursor cells are produced. These cells are essential to the immune system. Furthermore, it is from these precursor cells, also called stem cells, that specialized cells such as skin or blood are regenerated in normal cycles of repair. Moreover, stem cells from bone marrow can now be cultivated and are routinely used in a variety of medical treatments.

Practice 4

Skim the title, headings, and passage to identify the topic, thesis, main ideas, and key words that repeat throughout. Do this as quickly as you can. Do not write them down. Just keep the information in your mind as a preview. Check your answers with those in the Answer Key on page 297.

Stonehenge

Arguably, the most recognizable megalithic monument in Europe is Stonehenge on the Salisbury Plain in southern England. Like other henges, Stonehenge is a circular arrangement of large stones, surrounded by a ditch. Although it has not been determined exactly who was responsible for the construction of Stonehenge, it was probably built in stages over several hundred years, beginning about 3,000 B.C.E.

The last henge, completed about 1,500 B.C.E., was designed as a post and lintel structure in concentric circles. The outer ring, which is almost 100 feet in diameter, is constructed of huge sandstone or sarsen megaliths with smaller volcanic bluestones from Wales forming the inner ring. Inside the bluestones, a semicircle opens to a long avenue marked by uprights on an axis to the east identified by the Heel Stone, a large stone with a pointed top. A person standing at the center can view at the Heel Stone the exact spot where the sun rises at the summer solstice, the longest day of the year. It is assumed that Stonehenge must have been some kind of solar calendar or an early observatory.

Clearly, the most intriguing question involves the engineering methods that allowed early man to build such a remarkable structure. The gigantic stones in the outer ring are 22 feet high and weigh as much as 50 tons each. They have been traced to Marlborough Downs, 20 miles north of Stonehenge. Although most of the land is flat, at Redhorn Hill, the steepest part of the road, modern estimates suggest that more than 600 men would have been required to transport one of the stones uphill. The raising of each lintel, a stone beam that rests on top of two vertical stone posts, would also have required heroic effort. Based on the use of sledges, rafts, pulleys, and other machines available to Neolithic humans, work studies calculate that 30 million hours of labor would be the minimum to quarry the stone, transport it to the site, and complete the arrangement of the posts and lintels for the monument. The positioning of the stones to place the rising sun exactly over the Heel Stone during the solstice is also impressive.

Reading Strategy 5

Answer vocabulary questions quickly

You do not have to read the passage to answer most of the vocabulary questions on the iBT and the ITP TOEFL. First read the shaded vocabulary word in the passage, and then read the four answer choices in the question. You can probably identify the synonym without referring to the reading passage. This will save time. There are usually three or more vocabulary questions for each passage. By using this strategy, you will have more time to answer the other questions.

Example Questions

The word "accumulation" in the passage is closest in meaning to

- (A) decrease
- ● collection
- (C) pattern
- (D) change

The word "ordinary" in the passage is closest in meaning to

- ● common
- (B) old
- (C) large
- (D) heavy

Example Passage

Sedimentary Rocks

Sedimentary rocks, which are formed by the accumulation of sediments, have traditionally been classified into three basic types. The first type, *Clastic* sedimentary rocks, includes breccias conglomerate, sandstone, and shale, which are formed from mechanical weathering, a process that breaks down rock into smaller particles during periods of freezing and thawing, landmass uplift, expansion and contraction from the sun or fire, the action of animals, or other means that do not have a direct effect on the rock's chemistry. They are composed mostly of quartz and other ordinary minerals and have usually been transported and deposited at a distance from their formation.

The second type, *Chemical* sedimentary rocks such as rock salt and some limestone form when dissolved materials precipitate from a solution . . .

Practice 5

First, find the vocabulary word in the question. Then read the four possible answers and choose the synonym. You probably won't need to refer to the reading passage. Check your answer in the Answer Key on page 298.

Questions

1. The word "reliably" in the passage is closest in meaning to
   - Ⓐ generally
   - Ⓑ partially
   - Ⓒ finally
   - Ⓓ dependably

2. The word "precise" in the passage is closest in meaning to
   - Ⓐ correct
   - Ⓑ useful
   - Ⓒ ready
   - Ⓓ short

Selective Breeding

For centuries farmers have been experimenting with crops and animals to change their genetic makeup with a view to improving their traits. By selective breeding, they have been able to achieve results in offspring. Among the most successful achievements in selective breeding is corn. Hybrid varieties reliably produce full, sweet kernels compared with the original wild corn that yielded only a few kernels per stalk.

The problem with selective breeding is that it requires a long time to accomplish the results and they are not always precise. Recently, the scientific community has discovered how to accelerate and improve the process through biotechnology . . .

In order to answer vocabulary items quickly, you will need to recognize synonyms for academic vocabulary. One of the best resources for an academic vocabulary list is the Academic Word List (AWL) developed by Averil Coxhead at Victoria University of Wellington, New Zealand. The list contains 570 word families that appear with greatest frequency in a broad range of academic texts. Therefore, they are the words that you need to know both for the TOEFL and for your studies at English-language colleges and universities.

The 570 words are divided into 10 sublists, ordered so that those in the first sublist include the most frequent words and those in the last sublist the least frequent.

Test yourself with the most frequent list. As quickly as you can, write a synonym beside the word. Then check your answers with those in the Answer Key on page 298.

analyze	________	define	________
approach	________	derive	________
area	________	distribute	________
assess	________	economy	________
assume	________	environment	________
authority	________	establish	________
available	________	estimate	________
benefit	________	evident	________
concept	________	factor	________
consist	________	finance	________
constitute	________	formula	________
context	________	function	________
contract	________	income	________
data	________	indicate	________

individual ____________________

interpret ____________________

involve ____________________

issue ____________________

labor ____________________

legal ____________________

legislate ____________________

major ____________________

method ____________________

occur ____________________

percent ____________________

period ____________________

policy ____________________

principle ____________________

proceed ____________________

process ____________________

require ____________________

research ____________________

respond ____________________

role ____________________

section ____________________

sector ____________________

significant ____________________

similar ____________________

source ____________________

specific ____________________

structure ____________________

theory ____________________

vary ____________________

To study all of the lists, I recommend the following site:

http://simple.wiktionary.org/wiki/Wiktionary:Academic_word_list.

★★★**Bonus:** iBT Listening, Speaking, Writing
Academic vocabulary is found in all sections of the iBT as well as in the Reading sections of the iBT and the ITP.

Reading Strategy 6

Rely on glossaries for technical vocabulary

You do NOT need to know technical vocabulary in order to understand the passages on the Reading section. If a technical word is used in the iBT, the word is highlighted in blue. Click on the word to see a definition in a pop-up glossary. If a technical word is used in the ITP, you will find the definition in a glossary at the end of the passage.

Example Glossary

Process by which two atomic nuclei combine to create a single, more massive nucleus

A star in the last phase with a low surface temperature and a large diameter

Example Passage

Red Giants

Medium mass stars like the Sun fuse hydrogen within their cores into helium, thereby creating outward pressure. Gravity balances the pressure, retaining the size and shape of the star. It is estimated that in about another five billion years, the Sun will have used up all of the hydrogen in its core, at which point the pressure will be less than the force of gravity, and the star will begin to collapse. During this stage of compaction, a star begins to heat up and fuses the remaining hydrogen in the shell around the core. The Sun's burning shell of hydrogen will expand, burning the outer layers of the star.

When this occurs, the Sun will be transformed into a **red giant**, expanding so far out into the solar system that it will totally envelop Mercury. Finally, as the Sun expands, the heat will disperse, cooling the overall temperature; however, the core will increase until it is hot enough to fuse the helium nuclei into carbon, nitrogen, and oxygen, with iron at the very center. Iron cannot be used as fuel . . .

Practice 6

First, find the glossary definition for the technical word or phrase in the sentence. Then, rewrite the sentence without using technical words. Check your answers with those in the Answer Key on page 299.

Sentences

> ***Polyandry***
> The marriage of one woman to two men.

1. Although rare, polyandry may occur when the two men involved are brothers.

Although rare, the marriage marrying of one woman to two men may occur when the two men involved are brothers.

> ***Enzymes***
> Proteins or protein-based molecules that increase a chemical reaction in an organism.

2. Chemical digestion, which is carried on by certain enzymes, reduces the size of the particles of food in the stomach.

Chemical digestion, which is carried on by certain protein-based

> ***Ziggurats***
> A temple with a tower, often shaped like a terraced pyramid.

3. Sumerian art before 4,000 B.C.E. centered around ziggurats and sculptures of their rulers.

Sumerian art before 4,000 BCE, centered around a temples with a towers & sculptured pyramid.

> ***Incorporeal property***
> Intangible property that has value, but no physical substance.

4. Incorporeal property such as magical rituals or medicinal formulas are usually transferred before death because they require a period of learning.

> ***Transpiration***
> The loss of water by evaporation.

5. It is mainly through the leaves of plants that transpiration occurs.

> ***Amygdala***
> Part of the limbic system located in the temporal lobes of the brain.

6. The response in the left amygdala was greater when presented with a fearful stimulus as opposed to a pleasant stimulus.

> ***Lexicography***
> The compilation and editing of a dictionary.

7. It was during the Middle Ages that the science of lexicography came into existence.

Gymnosperms
A plant with seeds that are unprotected by an ovary.

8. By far, the largest living group of gymnosperms are pine trees.

____

____

____

Anomie
A breakdown in social standards in an individual or group.

9. Durkheim introduced the concept of anomie in the late 1800s to explain the response by society to the Industrial Revolution.

____

____

____

Disequilibrium
Loss of stability due to imbalance in supply and demand.

10. In spite of arguments to the contrary, altering the value of currency can cause disequilibrium.

____

____

____

Reading Strategy 7

Find pronoun references

When you see a pronoun that is shaded in the passage, try to identify the noun or phrase to which it refers. Notice that most of the time the noun or phrase will appear before the pronoun. Look for words and phrases before the pronoun in the same sentence or in preceding sentences. Be sure that your choice agrees with the pronoun in number (singular or plural) and gender (masculine or feminine). You can eliminate incorrect choices because they do not agree in number and gender. Watch for the common pronouns listed below.

he	him	his	this	one	who
she	her	her	that	some	which
it	it	its	these	a few—several	that
they	them	their	those	many	

EXAMPLE PASSAGE

Standardized Measurements

Standardized measurements are so basic and necessary that it is difficult to imagine a world without them, but, just like everything else, someone had to come up with the idea. Archeologists believe that some of the stones that they have found in Babylonia may have been standard weight measurements; a few are so accurate that they could be used today. In addition to stones, the Egyptians and the Greeks used a wheat seed as the smallest unit of weight, and the Arabs used a small bean called a *karob*. However, after Galileo began to observe the pendulum at the beginning of the 1600s, several scientists became interested in the length of the pendulum, and they chose the length that beats the second as a natural unit of measure.

Inspired by the one-second swing, a French cleric, Gabriel Mouton, devised a system of measurement using the size of the Earth, dividing the multiples into decimals. He proposed this system in 1670. Although the terminology was different, his concept of a universal unit was brought to fruition at the end of the 1700s, when scientific experimentation was exploding and researchers realized that precise measures were essential to their studies. By 1790, the French Academy of Sciences had created a system that was both simple and scientific, an advance that greatly contributed to the industrial revolution. The basic unit of length was set as a fraction of the circumference of the Earth and measures for weight, volume, and mass and for all other units were derived from it. Larger and smaller multiples were calculated by multiplying or dividing the basic units by 10, a feature that made using the system very easy because the decimal point would be moved to determine larger or smaller values. The Academy designated the term *metre* to the unit of length, a word derived from the Greek *metron*, which means a "measure." By 1900, approximately 35 countries had adopted the metric system, and, of course, now it is the preferred system of weights and measures for international trade and commerce as well as for scientific investigations.

Practice 7

First underline each of the pronouns in the passage. Then find the noun or noun phrase to which it refers, draw a box around it, and draw an arrow to the reference. Check your answers with those in the Answer Key on page 300.

The Missouri Compromise

The Missouri Territory was the first to be organized from lands acquired in the Louisiana Purchase. At the time that Missouri petitioned for statehood in 1819, Senate membership was evenly divided between states that were designated as slave-holders and those that were free. Clearly, the admission of a new state would destroy the balance, providing a voting majority for one side or the other. Although the Civil War was still forty years away, slavery was already a hotly contested topic in Congress with about equal numbers in favor and opposed to it. Since slavery had already been established in the Missouri Territory, it appeared that the scale would tip in favor of slave states.

During the long debate, the Territory of Massachusetts applied for admission as the state of Maine. Quick to recognize a way to settle the issue peacefully, Speaker of the House Henry Clay proposed a solution that has come to be called "The Missouri Compromise." Combining the two petitions for statehood, he fashioned a bill that allowed the Senate to retain equal numbers of Senators from slave and free states. Missouri would be admitted as a slave state at the same time that Maine was admitted as a free state. As part of the bill, Clay introduced a plan to maintain the balance of power between the two factions of the Senate as future territories petitioned it for statehood. A provision that divided the Louisiana Purchase at the southern boundary of Missouri, that is, 36 degrees and 30 minutes latitude, allowed for slave states south of the boundary and free states north of the border—with the exception of the Missouri Territory—to enter as a slave state because of its prior status.

Thomas Jefferson, now an aging statesman, saw the potential for a division in the union based on regionalism and the slavery question. By 1850, his prediction was very much a reality. Southern states threatened secession, and a second compromise was necessary to avoid it. Again brokered by Henry Clay with assistance from Stephen Douglas, it was a complicated effort consisting of five separate bills to deal with the land in the Southwestern Territories that were beginning to organize for statehood, all of which were south of the thirty-sixth parallel. The plan, commonly referred to as the Second Missouri Compromise, took four years to complete and was successful for less than a decade in forestalling the Civil War.

Reading Strategy 8

Use the screen and the arrows to locate answers

Most of the time, you will find the answer to the question on the left side of the screen at the point in the passage that is visible on the right side of the screen. You will not need to scroll down to another screen. You will also see an arrow at the beginning of the paragraph in the passage when the question refers to that specific paragraph number. Use the screens and arrows efficiently to find the answers quickly.

EXAMPLE QUESTION

According to Paragraph 1, which area suffered from the first cases of plague?

Ⓐ Africa

● Asia

Ⓒ Europe

Ⓓ North America

Paragraph 1 is marked with an arrow [➔]

EXAMPLE PASSAGE

The Plague

➔ Often referred to as the Black Death, the plague that savaged both Asia and Europe in the mid-14th century was probably a strain of bubonic plague and pneumonia. It is believed that the pandemic, which claimed more than 25 million lives in Asia and another 25 million in Europe, probably originated in China and was brought into Europe by traders, probably through Sicily. From there it spread to other Mediterranean ports in Italy and then through North Africa and Spain to France and England. Of course, urban areas with dense populations were more likely to be struck, but even then, it was uncertain where it would strike. For example, Milan had few instances of plague while Tuscany was devastated.

It is estimated that almost one third of the population of Europe died from the plague during a three-year period when the disease was rampant, and almost half of the people in England were victims of plague. Historians record that almost a thousand English villages were totally destroyed by disease. So what were the long-term effects of this devastation? Clearly, trade was affected, but probably even more important was the effect on agriculture. A shortage of workers among the peasant class in Europe caused land holders to compete for their services in order to cultivate the land. It was at this time that wages for work was instituted for peasants and craftsmen, a reform that began to blur the lines between classes in society. The cost of food and goods increased with the increase in the price of labor, but ultimately, in spite of a good deal of social unrest, workers benefited from the higher wages, merchants profited from the higher prices, and even landowners benefited from the higher land values . . .

Practice 8

First, read the question. Then use the screen and arrow to find the location of the answer in the passage. Check your answers with those in the Answer Key on page 301.

QUESTIONS

According to Paragraph 3, which of the following activities is included in imaginative play?

Ⓐ Playing peek-a-boo

Ⓑ Running and jumping

Ⓒ Building with blocks

Ⓓ Making up fantasies

Paragraph 3 is marked with an arrow [➔]

According to paragraph 6, why is play important in childhood?

Ⓐ It teaches rules for games.

Ⓑ It allows children to explore the arts.

Ⓒ It develops reasoning skills.

Ⓓ It helps children choose friends.

Paragraph 6 is marked with an arrow [➔]

Types of Play

Children engage in different types of play, and although it is helpful to categorize them, it should be understood that there is a great deal of overlap among them. Besides, several different methods of classification have been devised, a circumstance which adds to the confusion. However, it is useful to classify the major types that seem to appear in most of the lists.

The first type that is commonly identified is *sensorimotor play*, which begins in infancy and includes motor activities such as crawling, running, jumping, waving, or playing peek-a-boo. Sensorimotor play also includes opportunities for children to manipulate objects such as a rattle or a ball and allows them to enjoy physical stimulation as they explore the environment. These normal activities are easily observed as young children interact with people and objects at a very early stage of life.

➔ Another type of play that most psychologists have identified is *imaginative play*, which usually involves make-believe situations. Children pretend to be someone else or imagine an activity or place that is part of a fantasy world. Daydreaming is one form of imaginative play, but often children actually create involved scenarios and improvised dialogues. Since this happens after children are old enough to have memories, most adults recall playing "Make Believe."

Constructive play, also called *creative play,* is a third type. Examples include making music, either with real or toy instruments, creating art such as drawing, painting, molding with clay, building with blocks or other materials, as well as cooking and completing simple sewing projects. Of course, this can be structured in a classroom setting, or it can be a spontaneous event.

Finally, the fourth type of play, *cooperative play,* involves more structured activities which require interaction with other children in socially acceptable ways. One of the highest levels of cooperative play consists of games with rules, including not only sports like baseball, basketball, and soccer but also board games like checkers and Monopoly, or card games at varying levels of complexity.

➔ So what does this mean? The different types of play, in addition to teaching children social skills, also provide an important way to build cognitive skills. Children who are allowed to experience a wide range of play situations have demonstrated that they are more able to respond to unique or unfamiliar situations in real life. They react more acceptably in social settings and form more appropriate relationships.

Reading Strategy 9

Look for clues to scan for information

When you read a question that asks *who*, *where*, *when*, *how much*, or *what*, you should scan for details. You will know what you are looking for and you can move your eyes quickly through the passage to find it. Clues that will help you locate the answers to detail questions are identified below.

Capital Letters	***Numbers***	***Key Words***
Who? Where?	When? How much? How long?	What? Why? How?
To find the names of people and places	To find sums of money or length of time	To find terms and key words

Example Questions

What is an NPO?
Look for the key word

Who was the founder of the NPO that provided the original donation for the public library system in the United States?
Look for capital letters

When was the Carnegie Corporation of New York established?
Look for numbers

How much is the endowment for the Bill and Melinda Gates Foundation?
Look for numbers

Where is the Wellcome Trust located?
Look for capital letters

How is an NPO like a for-profit business?
Look for key words

What is the primary difference between an NPO and a for-profit company?
Look for key words

Example Passage

A Nonprofit Organization

A nonprofit organization, also called a not-for-profit organization or an NPO, is an organization that uses its surplus funds to achieve goals related to its mission instead of distributing dividends to owners or shareholders. Some government agencies are considered NPOs, but in many countries they are classified separately. For the most part, NPOs are charities, and as such, they are tax exempt.

Founded by Andrew Carnegie in 1911, the Carnegie Corporation of New York is still one of the most influential NPOs in the United States. During the first two years, Carnegie gave the Corporation $125 million, and acted as the president and trustee, overseeing charitable donations, notably including $43 million for public libraries, which he referred to as "the universities of the people."

More recently, the wealthiest NPO in the United States is the Bill and Melinda Gates Foundation, which has an endowment of $38 billion. Committed to healthcare, education, and access to technology, the foundation provides grants to many international organizations. In Great Britain, the Wellcome Trust, with an endowment of £14 billion, is a non-governmental trust with a goal of supporting biomedical research and public education in the areas of science and technology.

In addition to large NPOs, millions of smaller NPOs provide social services to local communities. One confusing aspect of NPOs is the fact that, like for-profit businesses, it is expected that NPOs will earn a profit, referred to as a surplus; however, unlike a for-profit company, an NPO must retain the funds for expansion and future projects, without benefit to interested individuals. Although some NPOs rely on unpaid volunteers, others hire and pay management personnel. The crucial difference is not whether management works for compensation but whether interested investors receive a return on their investment. If they do, the organization is considered a for-profit business.

Although it is not specifically restricted to NPOs, the .org designation on the Internet is commonly used by the nonprofits as compared with the .com designation generally reserved for companies that are classified as for-profit.

Practice 9

First, read the question. Then determine whether you are looking for capital letters, numbers, or key words. Use your scanning strategies to find the location of the answer in the passage and underline it. Check your answers with those in the Answer Key on page 302.

QUESTIONS

What is different about the skeleton of a seahorse?

How do seahorses swim?

When was research data made available about the seahorse's head?

Where are researchers studying the shape of the seahorse?

Who is the investigator using biomechanical analysis to understand the evolution of the seahorse?

How long does it take for seahorses to hatch?

How many young seahorses will survive to maturity?

Seahorses

Seahorses are unique in many ways. Although they are bony fish, their skeletons are composed of a series of plates arranged in rings around which a thin skin is stretched, and they do not have scales. They swim upright, which is also a distinct characteristic, and they are often found at rest because they are very poor swimmers compared to other fish. Seahorses move through the water by rapidly fluttering a dorsal fin and steering with a pectoral fin positioned behind the eyes. While they are resting, they curl their prehensile tails around a stationary object like a monkey might do in a tree to stabilize itself.

One of the most striking characteristics of seahorses is their shape. Their tiny heads closely resemble the horse for which they are named. Even the spikey crown looks like a small mane. In 2011, researchers at the University of Antwerp in Belgium released data that included several interesting clues about the horse-like appearance. Using biomechanical analysis, Dr. Sam Van Wassenbergh and his team learned that the seahorse can capture its prey of small shrimp at a greater distance than can the pipefish, which, along with the seahorse, descended from a common ancestor. According to Dr. Van Wassenbergh, the arched head of the seahorse enables it to bend its head and extend it in a snapping technique that allows it to achieve a larger striking distance. The team believes that natural selection would have favored this foraging behavior, and encouraged the horse-like head to develop.

Reproduction in seahorses is also distinct from most other animals. It is, in fact, the male that carries the eggs in a special egg pouch that serves as an incubator until the small seahorses hatch fully formed. During gestation, which lasts from two to four weeks, the female visits the male for a few minutes every morning, then swims away. The number of young varies dramatically from as few as 5 to as many as 1,500, only .5 percent of which will actually survive to adulthood.

★★**Bonus:** iBT Speaking, Writing
In addition to the Reading sections on the iBT and ITP, you can use clues to scan for information in the reading passages in the Speaking and Writing sections of the iBT.

Reading Strategy 10

Put check marks in your notes to identify exceptions

Some questions on the Reading section will ask you to identify an answer choice that is NOT mentioned in the passage. When you scan the reading passage to look for exceptions, it is important to keep track of and eliminate the answer choices that you find in the passage. You cannot do this on the iBT screen, but you can use your note paper. Write ABCD in your notes. Then, when you find an answer choice in the passage, put a check mark beside the letter that corresponds to the answer choice that you have found. When you have check marks beside three of the answer choices, the choice remaining is the exception. Use this system to keep from getting confused because you probably won't find the answer choices in the passage in the same order in which you see them presented in the question.

Example Question

Which of the following types of moraines was NOT mentioned in the passage?

Ⓐ An end moraine

Ⓑ A lateral moraine

● A recessional moraine

Ⓓ A ground moraine

Example Notes

A ✓
B ✓
C
D ✓

Example Question

According to the passage, moraines are found in all of the following locations EXCEPT

Ⓐ Under the path of a moving glacier

Ⓑ At the end of the path of a glacier

Ⓒ Along the sides of the path of a glacier

● Behind the path of a glacier

Example Notes

A ✓
B ✓
C ✓
D

Example Passage

Moraines

A moraine is a formation of rock and soil that is moved and shaped by a glacier. Some formations are evident on a glacier that is still moving; however, others have been transported and left by glaciers that have receded or have now disappeared. The shape and location of the particles, called till deposits, are used to classify the type of moraine observed. Although as many as eight distinct types of moraines have been identified and described, only four are common to most glacial formations.

A terminal moraine, also called an end moraine because it occurs at the end of a glacier, shows the nature of the glacier's movement. If the glacier has stopped advancing, then the terminal moraine marks the end of the glacier. The longer the glacier remains in place, the more rock and debris will pile up. On the other hand, if the glacier is receding, then smaller piles or ridges will be seen along the path of retreat. Occasionally, terminal moraines are damaged by erosion and may be difficult to identify.

When a glacier moves through rocky terrain, the pressure and extreme cold temperatures break large boulders of the rock face, which are thrown to the edges of the glacier as it continues to move. A lateral moraine appears at the sides or margins of a glacier. When the glacier melts, the lateral moraine forms large ridges that mark the edges of the glacial path.

A mixture of rock and debris that is deposited underneath a glacier creates a ground moraine. As such, the distribution is usually regular along the path that the glacier once flowed, and can sometimes be deposited between two lateral moraines, although it consists of smaller rocks because they have been ground and broken by the weight and pressure of the moving glacier.

When two glaciers meet and merge, medial moraines are formed. Rocks and debris pile up, forming a new ridge, which is now in the middle of a new, larger glacier. If the new glacier melts, then the medial moraine is exposed in the center of its former ice field. These medial moraines are evidence of the formation of a . . .

Practice 10

First, read the question. Then, quickly write the letters of the four answer choices in your notes. Using your scanning skills, try to find the choices in the passage and put a check mark beside the letter in your notes. Identify the answer that is NOT in the passage. Check your answers with those in the Answer Key on page 303.

Question

According to the passage, all of the following stimuli were used in Pavlov's experiments EXCEPT

Ⓐ Footsteps

Ⓑ A shock

Ⓒ A lab coat

Ⓓ A light

Notes

A ✓
B
C ✓
D ✓

Question

According to the passage, classical conditioning has been useful in all of the following EXCEPT

Ⓐ Treatments for anxiety

Ⓑ Modern research

Ⓒ Accidents

Ⓓ Advertising campaigns

Notes

A ✓
B ✓
C
D ✓

Classical Conditioning

Although the phenomenon had been observed prior to Pavlov's investigations, he was the first to appreciate the significance of *conditioned response*. An accidental discovery, conditioned response was observed while Pavlov was studying the gastric secretions of dogs in his laboratory. He noted that the sound of his associate's footsteps caused the dogs to salivate even before the food was brought in. Later, he was able to use other stimuli to evoke the same response. For example, by presenting a light followed by food, Pavlov could condition the dogs to salivate after the light was presented and before the food was offered. In this case, the salivation was an *unconditioned response* to food, the light was the *conditioned stimulus*, and the salivation was the *conditioned response* to the light.

Repeating the experiment in numerous variations, Pavlov determined that he could replicate the results with many neutral stimuli including a white lab coat, a bell, and a tuning fork as well as a light. By pairing the neutral stimuli with food, the dogs would salivate before the food appeared. Ultimately, he could elicit salivation by presenting only the neutral stimulus, without the presence of the food.

In these experiments, Pavlov established the basic terminology and methodology that continues to be used in modern classical conditioning experiments. The discovery that environmental events with no relationship to a given response could, through experience, cause the response through association, was a breakthrough in behavioral psychology.

Animals, like Pavlov's dogs, tend to associate stimuli with survival. Consequently, the use of food in animal experiments would be logical; however, human subjects may associate unrelated stimuli with complex emotions. People with unusual fears or phobias may have had an emotionally charged experience with which they have associated a harmless object. In that case, it would be appropriate to reverse the conditioned response. Pavlov called this reverse conditioning *extinction*, a process that is still very effective in treating anxiety.

Because conditioning is so basic to learned human behavior, commercial advertising has exploited Pavlov's research in order to influence purchasing behavior. Effective commercials often train the public to associate a product with a pleasant or desirable stimulus that evokes a positive response. In this case, the public has an unconditioned response to the product; the conditioned stimulus is a beautiful woman, a luxury lifestyle, or some other desirable image, and the purchase is a conditioned response.

Reading Strategy 11

Eliminate minor points in summaries

On the last question for each passage, you will see a sentence that expresses the main idea of the passage. Then you will be asked to summarize the passage by choosing three major points from six or more choices. When you look at the answer choices, you will be able to identify two types of incorrect choices—ideas that are NOT in the passage and minor points. Examples and details are minor points that support the major points. Eliminating the minor points will help you find the major points for the summary.

Example Main Idea

The teepee, widely used as a shelter by many tribes, had several advantages for life on the plains of North America.

Major Point
The teepee was easy to construct, disassemble, pack, and transport when the tribe moved to a new area.

> **Minor point**
> Poles, ropes, and skins could be folded and strapped to pack horses or loaded on horse-drawn sleds.

Major point
The animal skin or birch bark materials made a teepee warm in winter, dry in spring, and cool in summer.

> **Minor point**
> Old style teepees were made of buffalo hides and blankets but later structures used heavy canvas.

Major point
The unique opening at the top and smoke holes at the sides allowed the women to cook inside the shelter.

> **Minor point**
> Smoke from the fire in the middle of the teepee could rise and exit through the hole at the top.

Practice 11

First, read the introductory sentence that presents the main idea for the passage. Next, read the six possible answers and choose the three major points that correspond to the introductory sentence. Then find the minor point that is an example or detail for each major point. Organize the sentences and rewrite them as an outline for a short reading passage. Check your answers with those in the Answer Key on page 304.

1. **Noam Chomsky proposed the language acquisition device (LAD) as an explanation for the learning of native languages by young children.**

   The theory assumes that children are born with an innate facility for acquiring any language.

   First language learning occurs too rapidly for such a complex system unless children have the LAD.

   The LAD scans for input signals that allow children to focus and learn a specific language.

   Word order for sentences in languages is restricted to a small number of possibilities.

   Universal grammar, common to all languages, is present in the brains of human infants at birth.

   Children do not simply repeat sentences that they hear from limited, often fragmented adult input.

2. **The Hubble Telescope has provided invaluable information that has virtually revolutionized the field of astronomy.**

   Two planets were found orbiting around a pulsar in a neighboring solar system.

   Hundreds of planets have been discovered in regions beyond our solar system.

   Scientists have found an unexplained force called dark energy, which is accelerating the expansion of the universe.

   The expansion of the universe seems to be speeding up instead of slowing down.

   Galaxies vary in shape, color, and size, and their age goes back almost to the inception of the universe.

   The Hubble Deep Field reveals galaxies farther away than we have ever been able to see before.

3. **Humans have been adorning themselves with jewelry from ancient times, using local materials.**

   Ancient Western cultures preferred gold from as early as 3,000 B.C.E. for rings, earrings, and necklaces.

   Beautiful necklaces and earrings have been found in burial sites in Egypt, Greece, and Rome.

   In China, craftsmen used silver, which they enameled to create necklaces, bracelets, rings, and anklets.

   Prehistoric peoples made jewelry of leather or grass, which was strung with shells, stones, or bones.

   Jewelry designs by Chinese artisans included flowers, turtles, birds, and dragons.

   Some of the jewelry has been preserved in caves and in buried sites of ancient cities.

4. **Bird nests built in trees are generally categorized on the basis of their style.**

   Mud nests must dry out before each new layer is added to the nest or they would collapse.

   While some birds can complete a cup nest in one day, others require up to two weeks to construct their nests.

   Birds take advantage of natural cavities in trees, enlarging and shaping them by chiseling or chewing with strong beaks.

   Woodpeckers use holes that have been excavated by insects, and improve them by pecking out additional spaces.

   A nest situated in the fork of a tree is typically built with grass, twigs, and other materials shaped into the shape of a cup.

   Some nests are constructed of mud mixed with the bird's saliva, shaped, and allowed to dry against the tree trunk or a large branch.

5. **The work of e.e. cummings does not conform to the usual conventions of poetry and language.**

Adjectives such as "mud-luscious" and "watersmooth" create images unlike that of any other poet.

The author preferred to use lower case letters for his own initials and last name on his poems.

The absence of punctuation was one of the ways that the poet departed from the traditional rules.

Cummings's inventive formations of compound words make his poems both surprising and unique.

Cummings did not always capitalize the personal and place names as required by the rules for using capital letters.

Eccentric punctuation was characteristic of Cummings's unique literary style.

★★**Bonus:** iBT Speaking, Writing
In addition to the Reading sections on the iBT and ITP, you will need to use major points to summarize information on the iBT in the Speaking section and the Writing section.

Reading Strategy 12

Draw logical conclusions

Sometimes you will be asked to make an inference based on evidence in the passage. When information is not directly stated, you must draw a logical conclusion by recognizing the relationship between several facts. For example, in the passage you may learn that all birds have beaks. You may also learn that a Kiwi is a bird. A logical conclusion is that a Kiwi must have a beak. When you read, think about how the facts are connected.

Example Questions

Which of the following can be inferred about elk?

Ⓐ They compete with wolves for prey.

Ⓑ They have the same diet as cougars and bears.

Ⓒ They have declined because of disease.

● They are a major food source for wolves.

Because *other* predators compete for elk, it must be concluded that *wolves* are also a predator of elk, and elk are a major food source for wolves.

It can be inferred from the passage that mange is a

Ⓐ young wolf

● disease

Ⓒ predator

Ⓓ food

Because *disease* is a factor in the decline of wolf packs, and pups are vulnerable to and killed by *mange*, it must be concluded that mange is a disease.

Example Passage

Wolves

Although the wolf population has experienced a recent decline in Yellowstone National Park, a number of ecologists believe that the wolves are simply maintaining equilibrium with the elk population, which has also declined during the same time frame. Other predators compete for elk, including cougars and bears; therefore, when cougar and bear populations increase or hunt more successfully, then wolves experience higher mortality rates as a natural consequence.

Furthermore, when any animal is not well nourished, its immune system is not strong, and it is more susceptible to disease, which can also be a factor in the decline of wolf packs. Pups are particularly vulnerable to mange, which can kill as many as 80 percent of the young, thus exposing the population to decline from one generation to the next. . . .

Practice 12

First, read the related sentences in each set. Based on the evidence from the sentences, write a logical conclusion. Check your answers with those in the Answer Key on page 306.

1. Drums were used in ancient Chinese culture for all ceremonial occasions. . . . Weddings and funerals were among the most common ceremonies.

2. Fossils are the remains or impressions of plants or animals, embedded in rock. . . . Petrified wood results when organic materials are replaced with minerals, thereby turning a tree into stone.

3. The great railroad hotels were built close to the railroad lines in the late nineteenth and early twentieth century at stops near landmarks of interest to tourists. . . . The Empress Hotel in Victoria is one of the most elegant railroad hotels in Canada.

4. Unlike ferns, mosses and liverworts lack complex vascular tissue. . . . The absence of an extensive vascular system restricts the size of plants.

5. Studies at Princeton University suggest that it takes most subjects one-tenth of one second to form an impression of a stranger's face. . . . Several minutes of exposure to the face does not change the initial impression, although subjects do report greater confidence about their judgments.

Reading Strategy 13

Study the first word or phrase in the insert sentence

When you are asked to choose where to insert a sentence in a passage, you should pay special attention to the first word or phrase in the sentence to be inserted. Nouns or pronouns may refer to nouns in previous sentences. Sometimes transition words or phrases make a logical connection between the insert sentence and the previous sentence. The most common transitions are listed below.

Contrast	***Addition***	***Example***	***Restatement***	***Order***
On the other hand	Furthermore	For example	In other words	First
On the contrary	Moreover	For instance		Next
In contrast	In addition			Then
				Finally

Example Questions

1. Where would the following sentence best fit in?

One such story is that of a great flood that covered the world and destroyed all but a few people.

Ancient myth in many distinct parts of the world consists of common stories. [INSERT SENTENCE] Furthermore, the construction of an ark or some other vessel figures prominently in most of these stories. For example, in Babylonia, the gods were angry because there were too many people on the Earth, and they caused a seven-day deluge, but Atrahasis and his family escaped on an ark with animals. According to Hindi myth, Vishnu, in the form of a fish, warned Satyaavrata of the impending flood in time for him to save his family and his animals on an ark. In contrast, the Mayan myths in Mexico recounted three floods, the last of which destroyed all but three people who escaped in a canoe. In Africa, Europe, the Far East, Australia, and North and South America the same story was repeated with regional variations. In other words, the story is basically the same, but the details reflect the local culture.

This insert sentence includes the phrase "One such story," which refers to the noun phrase "common stories" in the previous sentence.

2. Where would the following sentence best fit in?

   Furthermore, the construction of an ark or some other vessel figures prominently in most of these stories.

   Ancient myth in many distinct parts of the world consists of common stories. One such story is that of a great flood that covered the world and destroyed all but a few people. [INSERT SENTENCE] For example, in Babylonia, the gods were angry because there were too many people on the Earth, and they caused a seven-day deluge, but Atrahasis and his family escaped on an ark with animals. According to Hindi myth, Vishnu, in the form of a fish, warned Satyaavrata of the impending flood in time for him to save his family and his animals on an ark. In contrast, the Mayan myths in Mexico recounted three floods, the last of which destroyed all but three people who escaped in a canoe. In Africa, Europe, the Far East, Australia, and North and South America the same story was repeated with regional variations. In other words, the story is basically the same, but the details reflect the local culture.

   This insert sentence includes the transition word "Furthermore," which introduces the additional information about the "ark" in most of the "common stories" referred to in previous sentences.

3. Where would the following sentence best fit in?

   For example, in Babylonia, the gods were angry because there were too many people on the Earth, and they caused a seven-day deluge, but Atrahasis and his family escaped on an ark with animals.

   Ancient myth in many distinct parts of the world consists of common stories. One such story is that of a great flood that covered the world and destroyed all but a few people. Furthermore, the construction of an ark or some other vessel figures prominently in most of these stories. [INSERT SENTENCE] According to Hindi myth, Vishnu, in the form of a fish, warned Satyaavrata of the impending flood in time for him to save his family and his animals on an ark. In contrast, the Mayan myths in Mexico recounted three floods, the last of which destroyed all but three people who escaped in a canoe. In Africa, Europe, the Far East, Australia, and North and South America the same story was repeated with regional variations. In other words, the story is basically the same, but the details reflect the local culture.

   This insert sentence provides an example of the story of the "ark" referred to in the previous sentence.

4. Where would the following sentence best fit in?

   In contrast, the Mayan myths in Mexico recounted three floods, the last of which destroyed all but three people who escaped in a canoe.

   Ancient myth in many distinct parts of the world consists of common stories. One such story is that of a great flood that covered the world and destroyed all but a few people. Furthermore, the construction of an ark or some other vessel figures prominently in most of these stories. For example, in Babylonia, the gods were angry because there were too many people on the Earth, and they caused a seven-day deluge, but Atrahasis and his family escaped on an ark with animals. According to Hindi myth, Vishnu, in the form of a fish, warned Satyaavrata of the impending flood in time for him to save his family and his animals on an ark. [INSERT SENTENCE] In Africa, Europe, the Far East, Australia, and North and South America the same story was repeated with regional variations. In other words, the story is basically the same, but the details reflect the local culture.

   This insert sentence contrasts "three floods" with one "impending flood" in the previous sentence.

5. Where would the following sentence best fit in?

   In other words, the story is basically the same, but the details reflect the local culture.

   Ancient myth in many distinct parts of the world consists of common stories. One such story is that of a great flood that covered the world and destroyed all but a few people. Furthermore, the construction of an ark or some other vessel figures prominently in most of these stories. For example, in Babylonia, the gods were angry because there were too many people on the Earth, and they caused a seven-day deluge, but Atrahasis and his family escaped on an ark with animals. According to Hindi myth, Vishnu, in the form of a fish, warned Satyaavrata of the impending flood in time for him to save his family and his animals on an ark. In contrast, the Mayan myths in Mexico recounted three floods, the last of which destroyed all but three people who escaped in a canoe. In Africa, Europe, the Far East, Australia, and North and South America the same story was repeated with regional variations. [INSERT SENTENCE]

   This insert sentence restates the previous sentence.

The myth follows the same basic storyline in all of the variations. First, a good person is warned of the flood. Next he prepares by constructing an ark or a boat for his family and animals. Then the world is destroyed. Finally, the family begins life on Earth in an effort to create a better world.

Insert sentences would fit into the sequence of events, using the first word to find a logical connection.

Practice 13

First, read the related sentences in each set. Based on the first word or phrase in the second sentence, make a logical connection between the sentences. Check your answers with those in the Answer Key on page 307.

1. There is a great deal of controversy about how to distinguish a lake from a pond, especially among the international scientific community.

   In other words, an internationally accepted definition of a pond cannot be agreed upon.

2. Crater lakes are sometimes formed in extinct volcanic craters that fill up with fresh water more rapidly than they evaporate.

   Moreover, these fresh water lakes tend to have exceptional clarity because they lack inflowing stream water that would muddy them with sediment.

3. Near the end of the last Ice Age, glaciers began to retreat, leaving behind large deposits of ice in the depressions that they had gouged out, creating lakes as the ice melted.

   For example, in Minnesota, Lake Duluth was formed at the southern tip of Lake Superior, originally occupying a much larger area than Lake Superior does now.

4. A glacial lake is a lake that is created by a melted glacier, as for example, a kettle lake.

   In contrast, an oxbow lake is formed when a river meanders and becomes very curved.

5. Lakes formed by the impact of a meteor have distinct characteristics. First, they tend to be perfectly round. In addition, a raised rim often outlines the depression.

   Finally, the water is very pure, since it is fed by rain or snow.

Reading Strategy 14

Divide a complex sentence into two simple sentences

You will see complex sentences highlighted in the reading passages. You are to choose the answer choice that paraphrases the highlighted sentence. The paraphrase must not add or leave out important information. To make the best choice, first make two simple sentences from the complex sentence in the passage. Then compare the information in the simple sentences with information in the paraphrase sentences in the answer choices. Be sure to eliminate a choice that changes the meaning or a choice that includes only part of the information.

Example Sentences

Original sentence in the passage:
Many modern drugs are derived from plants that have been used as cures for thousands of years by herbalists.

Two simple sentences:
Many modern drugs are derived from plants.
Plants have been used as cures for thousands of years by herbalists.

Complex sentence with the same meaning:
For thousands of years, herbalists have used many of the same plants as those found in modern drugs to cure diseases.

Original sentence in the passage:
The African elephant can be distinguished from the Asian elephant by its larger ears, heavier body weight, and taller stature.

Three simple sentences:
The African elephant has larger ears than the Asian elephant.
The African elephant weighs more than the Asian elephant.
The African elephant is taller than the Asian elephant.

Complex sentence with the same meaning:
Compared with African elephants, Asian elephants are shorter, they weigh less, and their ears are not as large.

Original sentence in the passage:
The most widely spoken language In South America is not Spanish but Portuguese, even though it is the official language of only one of the countries on the continent.

Three simple sentences:
The most widely spoken language in South America is not Spanish.
The most widely spoken language in South America is Portuguese.
Portuguese is the official language of only one of the countries on the continent of South America.

Complex sentence with the same meaning:
The official language of only one of the countries in South America, Portuguese surpasses Spanish as the most widely spoken language on the continent.

Original sentence in the passage:
Although the invention of the Internet has been attributed to a number of individuals, the creation required the combined effort of many scientists, researchers, and agencies.

Two simple sentences:
The invention of the Internet has been attributed to a number of individuals.
The creation required the combined effort of many scientists, researchers, and agencies.

Complex sentence with the same meaning:
The creation of the Internet required the combined effort of many scientists, researchers, and agencies, even though the invention has been attributed to a number of individuals.

Original sentence in the passage:
The human body in Romanesque sculpture was characteristically elongated because the artist was expressing the spiritual form.

Two simple sentences:
The human body in Romanesque sculpture was characteristically elongated.
The artist was expressing the spiritual form.

Complex sentence with the same meaning:
The artist characteristically expressed the spiritual form of the human body by elongating it in Romanesque sculpture.

Original sentence in the passage:
The pancreas not only supports digestion by producing enzymes and digestive juices but also regulates blood sugar by releasing important hormones directly into the bloodstream.

Two simple sentences:
The pancreas supports digestion by producing enzymes and digestive juices.
The pancreas regulates blood sugar by releasing important hormones directly into the bloodstream.

Complex sentence with the same meaning:
The pancreas produces enzymes and digestive juices to support digestion and releases hormones into the bloodstream to regulate blood sugar.

Practice 14

First, read the simple sentences. Then, using the simple sentences, write a new complex sentence with the same meaning as the original sentences. Check your answers with those in the Answer Key on page 308.

1. The Mycenaean civilization fell.
   Villages began to unite in order to form strong trading centers.
   Greek city-states evolved.

   ____

   ____

   ____

2. The peer group in schools is significant in the social experience of a growing child.
   The family unit remains fundamental throughout the entire life span of an individual.

   ____

   ____

   ____

3. Bones are not considered a solid structure.
   They generally have a large cavity in the center to accommodate marrow.

   ____

   ____

   ____

4. Changes in technology have influenced the advertising industry dramatically in the past few decades.
   The pop-up ad on the Internet is an example of the influence.

   ____

   ____

   ____

5. Studies with human subjects have confirmed that eating 30 percent fewer calories can increase one's life span.
   The essential nutrients are included in the lower-calorie diet.

   ____

   ____

   ____

6. The first printing presses were inspired by screw presses.
   Screw presses were used in agricultural industries for pressing grapes for wine and olive oil seeds for oil.

   ____

   ____

   ____

7. To survive and thrive, coral reefs require water that is at least 20 degrees Celsius.
   Coral reefs require adequate sunlight.
   Coral reefs require clear water, less than 100 meters deep.

   ____

   ____

   ____

8. Art historians suggest that some important works were previously attributed to male artists.
   These important works may actually have been created by women.
   Women were unable to market their art under their own names.

   ____

   ____

   ____

9. Many successful authors have had their books rejected by publishers, including J.K. Rowling.
   J.K. Rowling was the creator of the Harry Potter series.
   The original manuscript of the Harry Potter series was refused by five publishers.

   ____

   ____

   ____

10. Songbirds are often prized for their beautiful coloring and their lovely songs.
    Songbirds are an aesthetic addition to the environment.
    Songbirds are a practical means to control insects and weeds without chemical pesticides.

    ____

    ____

    ____

★★★**Bonus:** iBT Listening, Speaking, Writing
Paraphrasing is a skill that you will need on every section of the iBT TOFEL.

Reading Strategy 15

Read with your eyes and your brain

Do you say the words quietly aloud while you are reading? Even if you can't actually hear the words, do you move your lips or "hear" the words in your head? Maybe you try to pronounce the difficult words. This habit, called *subvocalization*, will limit your reading speed to the rate at which you can speak—about 150 words per minute, which is much too slow to succeed on reading tests like the TOEFL. Two tricks will help you break the habit. Put a pencil in your mouth or put your finger over your lips. To read faster, you must stop reading with your mouth and your ears and start reading with your eyes and your brain.

EXAMPLE PASSAGE

Read without moving your lips.

Animal Markings

It is widely known that no two human fingerprints are exactly the same, but humans are not the only species to identify themselves as unique among the crowd. Animal markings on zebras also identify individuals as unique members of their species.

StripeSpotter, a joint effort of the University of Illinois and Princeton University, is a computer program that identifies animals by their markings. After an animal is captured in an image, the software can identify the animal from other images, even when the animal is part of a herd. This is a superior way to study individual animals because the animal does not have to be captured and marked or labeled with tags or computer chips. The old capture and release systems were invasive, time consuming, and expensive. Now, ecologists in the field can study animals from a distance by taking pictures of them with digital cameras and large numbers of individuals can be captured on film in a short time by one scientist with a relatively low-priced piece of equipment.

StripeSpotter uses a sample of the markings on an animal from a photograph and slices it into a series of horizontal bands which are encoded in a format similar to that of the barcodes on products in the supermarket. When a similar code is read in another photograph, the animal is identified. Like the previous capture and release systems for studying animals, StripeSpotter allows scientists to track migrations and document relationships and interactions among individuals in groups.

StripeSpotter is already accumulating data on zebras in Kenya with surprising accuracy. According to studies, the system can probably be modified for additional animals with distinctive markings. Data bases for giraffes and tigers are being field-tested, and preliminary work is being amassed to include penguins, rhinos, and turtles, all of which have been studied and tagged using older capture and release methods.

Practice 15

First answer the question on the left. Then practice your new habits by reading the passage on the right. Try not to move your mouth while you read. For this practice, you can put a pencil in your mouth or put your finger over your lips. Check your answer in the Answer Key on page 310.

QUESTION

What can you do to stop saying or hearing the words in your head?

League of Nations

The League of Nations was established at the end of World War One in a covenant agreed to in the Treaty of Versailles. After a long and devastating global war, the League was formed in an effort to bring stability to the world. The mission of the new organization was to promote international cooperation and to maintain world peace. The League was based in Geneva, Switzerland, because Switzerland had been a neutral nation during the war.

In the event of an international incident, it was agreed that member nations would not go to war before submitting their dispute to the League, which would assist the nations involved to discuss their differences in an orderly forum. In addition, the League could impose a warning to an aggressive nation with consequences in the form of sanctions if that nation did not cooperate. Economic sanctions in the most severe form would include a trade embargo by League members. When economic sanctions were imposed, theoretically, it would cause distress within the offending nation to such an extent that the population would be motivated to pressure the government to accept the League's terms before the economy was bankrupted. If economic sanctions failed, then the League could impose order through military force.

The obvious flaw in the organization was the fact that the League did not have an army and the only countries in the League with a military presence were Great Britain and France, both of which were weakened by the recent war effort. Neither Germany nor Russia was allowed to participate, and the United States chose not to join. Therefore, although the League had lofty ideals—to preserve world peace—there was little chance for it to succeed on a political level because it did not have a powerful enough presence. Nevertheless, the League did have some victories in the social arena. The Health Organization made progress in providing fresh water wells and eliminating leprosy as well as improving the status of women and attacking the issue of child labor worldwide. When the League is evaluated and criticized, it is important to remember that it was the League that began many of the social programs continued through the United Nations today.

★★★**Bonus:** iBT Listening, Speaking, Writing

★★**Bonus:** ITP Listening, Structure

Reading passages are included in both the Speaking and the Writing sections of the iBT. In addition, reading without subvocalizing is important when you are reading the answer choices on all parts of the TOEFL.

Reading Strategy 16

Focus on larger chunks of text

If you look at one word at a time, your eyes will have to stop and focus 800 times in an 800-word passage. This habit, called *fixation*, will really slow you down. If you expand your field of vision to four words, you will reduce the number of times that your eyes stop to 200, and that will improve your speed! The best way to begin expanding is to look for meaningful phrases. Then you will stop at comprehension points and your comprehension will improve also. When you take the iBT, try to focus on an entire line in the passage on the right side of the screen. Focus on the words in the middle and your eyes will see the entire line. This is called *peripheral vision*. You use it all the time when you drive a car or ride a bicycle or motorcycle. You can use it when you read, too.

EXAMPLE PASSAGE

Try to focus on an entire line.

Pennsylvania

Designed as a refuge for Quakers,
Pennsylvania was founded by William Penn,
a member of an important British family
who had converted to Quakerism in his youth.
The main tenent of the Quaker faith was
pacifism and the belief that the followers could
commune with God without the intervention
of institutions or clergy. In Great Britain and
other European nations, Quakers were seen
as a threat to the established order.

In 1681, Penn requested and received
a land grant from King Charles II
in payment of a substantial debt
that the king owed to Penn's late father.
An extremely generous charter
made Penn the world's largest landowner
apart from the royal crowns of Europe,
with holdings of 45,000 square miles.
Originally, Penn gave the colony the name
Sylvania, which means "forest" in Latin.
Ultimately, King Charles changed the name
to *Pennsylvania* to honor William's father.

When Penn offered full religious freedom
in addition to a representative assembly
and attractive terms for land,
he was able to interest settlers from Europe.
Freedom of worship was total,
a benefit that attracted not only Quakers but
also other persecuted minorities from Northern
Europe, including Amish, Catholics, Lutherans,
and Jews.

Penn traveled to America
to establish the colony.
He personally planned the city of Philadelphia.
In keeping with his peaceful intentions,
he called it the "City of Brotherly Love."
A few weeks after his arrival,
Penn called a delegation of local tribes
with whom he was cultivating relationships.
They met under a large elm tree
where he offered "good faith" and "goodwill."
The treaties that they signed were kept
by both sides for generations.
While the northern colonies defended
themselves from attacks by warring tribes,
no Quakers were ever killed by Indians.

The new government of Pennsylvania was
established to assure the rights
of its citizens and was preserved in the
"Framework of Government," which Penn wrote.
Initially influenced by John Locke,
Penn created a completely unique concept—
the idea that amendments allowed the law
to evolve as times changed.
It was a precursor to the Constitution
of the United States with its amendments,
a document that would be written
almost one hundred years later.

Practice 16

First answer the question on the left. Then practice your new habits by reading the passage on the right. Focus on larger chunks of text. For this practice only, you can mark some of the chunks to get used to finding meaningful phrases or you can try to focus on an entire line. Try to find 5–10 words for each chunk. Check your answers with those in the Answer Key on page 311.

Question

What will help you find larger chunks of text to focus on with your eyes?

Bipedalism

Bipedalism, that is, walking upright on two legs, is one of the characteristics of human behavior that distinguishes us from our closest relatives on the evolutionary chart—the chimpanzees, gorillas, and other nonhuman primates. Fossil evidence suggests that bipedalism occurred soon after the evolutionary divergence of human beings from apes, and is considered an important development, one that defined us as *Homo sapiens*.

Many theories have been put forward to explain bipedalism, but several of them are fairly far-fetched, and only about half of them have gained acceptance among the scientific community. One of the more popular theories assumes that walking on two legs made it possible for early humans to use their arms to reach for food and carry food. This theory argues that chimpanzees occasionally walk upright when they are grabbing food from overhead branches or carrying food for later consumption. Thus, bipedalism actually supported survival because it made it possible to gather food more efficiently or to carry food to others in the group.

Another theory, again citing evolutionary bipedal development as a survival mechanism, asserts that raising the head to a higher level while walking upright extended the field of vision, and that made it possible for human beings to see both predators and prey at a greater distance. Therefore, the humans that stood and walked upright would have had an advantage and natural selection may have favored them.

Yet another theory suggests that raising the body above the ground level helps in the dissipation of heat because more favorable breezes adjust the body temperature. Again the evidence for this theory relies on observation of nonhuman primates who walk upright in order to cool down.

Finally, there is the theory that stone tools and weapons were easier to use and more efficiently carried when human beings had freed their hands for that purpose. And this particular theory has a lot of support among anthropologists who associate the use of tools with the evolutionary leap from ape to human.

★★★**Bonus:** iBT Listening, Speaking, Writing

★★**Bonus:** ITP Listening, Structure

Reading passages are included in both the Speaking and the Writing sections of the iBT. In addition, reading without fixating on small chunks of text is important when you are reading the answer choices on all parts of the TOEFL.

Reading Strategy 17

Move forward, not backward on the page or screen

If you take two steps forward and one step backward, you won't go very fast. The same idea applies to reading. If you keep going back and rereading the same words, you won't go very fast. This habit, called *regression*, will slow you down and will also interfere with your comprehension. Sometimes you need to read a sentence more than one time in order to understand it, but if you are rereading many sentences in each passage, you should try to break this habit. To avoid rereading on paper, put your hand or a pencil over the line that you have just read and continue by sliding your hand or pencil down to hide each line after you have read it. To avoid rereading on screen, scroll down to hide the previous lines.

Example Passage

Slide a pencil down from the top to hide each line that you have read in the passage on the right.

Prohibition

The Eighteenth Amendment to the Constitution of the United States prohibited the manufacture, sale, or transportation of alcoholic beverages. The amendment was fueled in part by the Women's Temperance Movement, which had achieved greater influence as it became clear that women were going to win the right to vote and elected officials wanted to appear to be sympathetic to their cause. Although President Wilson had vetoed the National Prohibition Act, Congress passed it over the veto in January 1919. Initially, light beer and wine continued to be sold legally; however, with pressure from lobbyists, in October Congress ratified the Volstead Act, in which the legal definition of an alcoholic beverage at one-half of one percent alcohol made even the light beverages illegal.

The key issue surrounding Prohibition was the ability to enforce the law, which proved difficult especially in urban areas. Saloons and bars were shut down only to be replaced by secret clubs called "speakeasies" that were supplied by bootleggers, that is, people who smuggled or manufactured hard liquor. In addition, small stills were set up to make home brews of all kinds and quality. Whether the amount of alcohol actually consumed was reduced is questionable, but historians agree that Prohibition actually increased the activities of organized crime, previously involved in gambling and prostitution, and now heavily invested in illegal alcohol.

Prohibition became more and more unpopular, and by the mid 1920s, it was an issue that divided the country. Rural areas continued to support the law, whereas urban areas favored repealing it. As criminal activity and violence increased, more people became disillusioned. Law enforcement at local levels had always been minimal, and corruption among low-level and high-level officials was rampant. By 1933, public opinion was clearly on the side of repeal, encouraging Congress to propose the Twenty-first Amendment. It went into effect on December 15, having been ratified by state ratifying conventions, selected specifically for that purpose. It is noteworthy that this amendment is the only one passed for the express purpose of repealing a prior amendment to the Constitution.

Practice 17

First answer the questions on the left. Then practice your new habits by reading the passage on the right. Try to continue reading without going back to reread the same words. For this practice only, use a paper to cover previous lines in the passage. Check your answers with those in the Answer Key on page 312. Then practice reading passages on a computer screen.

Questions

How can you avoid rereading when you are looking at a page? At a screen?

Be sure to practice reading text on a computer screen to acquire this skill.

Musicology

Music history, also referred to as historical musicology is the study of composition, performance, and critique over time. For example, a historical study of music would include biographical information about each composer and the way that a specific piece relates to other works, the development of styles of music, the relationship of music to social life, and the techniques associated with performance.

In theory, music history could treat the study of any type of music, but in practice, it has heavily favored Western music, and in particular, classical music. This perspective is problematic for two reasons. First of all, it dismisses music from other cultures. Because classical music was developed in nineteenth century Europe, it is representative of Western culture. But Eastern music has a rich, but very different tradition, which is virtually absent from the field of music history. Another problem is that music history arbitrarily ignores popular music, even Western music that has been composed post nineteenth century. Furthermore, the vocabulary and therefore the way that we talk about music is restricted because classical music is different from many other types of music in different cultures and time periods.

Consequently, when we look outside of Western classical music, we tend to look through the prism of classical music and often judge other music as inferior because it does not match the model that has been set up. For instance, classical music has a long tradition of harmony, but rhythm is not a key feature. In contrast, popular music, which has a very simple melody but a complex rhythmic structure, may be viewed as less valuable because it doesn't fit the mold. Another issue is the absence of notation in many non-Western forms of music. The fact that the rise of classical music coincided with the practice of musical notation elevates music that has a written tradition. Jazz, for example, would be considered less important because it relies on improvisation instead of sheet music.

★★★**Bonus:** iBT Listening, Speaking, Writing

★★**Bonus:** IPT Listening, Structure

Reading passages are included in both the Speaking and the Writing sections of the iBT. In addition, reading without regressing is important when you are reading the answer choices on all parts of the TOEFL.

Reading Strategy 18

Time each passage

To complete the Reading section, you should spend 20 minutes reading each passage and answering all of the questions that correspond to it. You can hide or show the clock in the iBT while you are reading, but you should not spend much time checking the clock. Pace yourself by checking once at the end of each passage. You should have 20 minutes left on your timer when you begin the last passage on the Reading section. Remember that the short version includes three passages and the long version includes four passages.

EXAMPLE iBT

The TOEFL presents 12–14 questions for each passage. These examples show a four-passage test with 12 questions on each passage and a three-passage test with 12 questions on each passage.

BEGINNING OF FOUR-PASSAGE READING SECTION **80:00**

Test Passage 1
Question 12 Show Timer 60:00

Test Passage 2
Question 24 Show Timer 40:00

Test Passage 3
Question 36 Show Timer 20:00

Test Passage 4

BEGINNING OF THREE-PASSAGE READING SECTION **60:00**

Test Passage 1
Question 12 Show Timer 40:00

Test Passage 2
Question 24 Show Timer 20:00

Test Passage 3

Practice 18

First, set a timer for 10 minutes. Use your watch or the clock on your smartphone. Then read the passage on the right and answer the first 5 questions. Were you able to complete them in 10 minutes? Check your answers with those in the Answer Key on page 313.

TIMED PASSAGE

The Hudson River School

The Hudson River School was a movement of mid-19th-century American artists who created paintings of landscapes. Initially, they limited their subjects to locations near the Hudson River Valley and the surrounding areas, including, for example, the Catskill Mountains, the Adirondack Range, and the White Mountains. The school was not an academic institution, but rather, it was an informal association of friends who studied and traveled together. Some of them even shared studios in New York.

Although the origin of the name is obscure, the Hudson River School was used primarily as a disparaging designation because the aesthetic vision of the group was considered outdated by art critics who preferred the Barbizon School. Hudson River School paintings depicted a realistic, if somewhat idealized setting, often combining rural landscapes with rugged stretches of wilderness. To achieve this juxtaposition, it was often necessary to merge several scenes, some from extreme environments. Unlike the Barbizon painters, who insisted that painting should be done on site in the open air, the Hudson River School used pencil sketches or memory to provide a record for compositions that would be painted at a later time in a studio. The inspiration for both artistic movements was an appreciation for the disappearing wilderness and a belief in nature as a manifestation of God, a view that they shared with contemporary literary schools such as the Transcendentalists. It was probably a book, *Essay on the Nature and Principles of Taste,* by Archibald Alison, that provided the most unified inspiration for many painters in the Hudson River School. According to Alison, for the people who sought harmony with it, the unspoiled beauty of nature could serve as an inspiration for good moral character.

Perhaps the most famous painter that emerged from the original Hudson River School was Thomas Cole, who, in the 1820s, embarked on an steamship up the Hudson and then headed West into the Catskill Mountains, where he painted some of the first landscapes of the area. Born in England, Cole appreciated the more vibrant colors of the American autumn. Many of his works portray colorful foliage at the peak of the fall season. The New World with seemingly limitless natural resources and wonders inspired Cole and his followers.

TIMED QUESTIONS

1. The word "obscure" is closest in meaning to which of the following
   - Ⓐ accurate
   - Ⓑ unusual
   - Ⓒ vague
   - Ⓓ complex

2. The word "it" in the passage refers to
   - Ⓐ harmony
   - Ⓑ nature
   - Ⓒ inspiration
   - Ⓓ character

After Cole's death in 1848, a new generation of Hudson River School artists inherited the movement. In fact, it can be argued that the second generation surpassed the first, producing many of the best representative works of the school between 1855 and 1875. Many of the painters studied at the Dusseldorf Academy or were influenced by work at the Academy. Like those of the first generation, many of their paintings were landscapes of the Hudson River and surrounding areas, but the second generation also traveled widely to Europe, South America, the Middle East, and the American West, searching for inspirational landscapes. Returning to their studios, they worked from pencil sketches, painting small works in oil, which they later developed into larger paintings.

➔ A student of Cole, Frederic Church, was known for enormous canvases such as the ten-foot wide *Heart of the Andes* as well as a large-scale landscape of Niagara Falls. Another of the leading members of the Hudson River School during the later years was Sanford Robinson Gilford. Originally trained as a medical doctor, Gilford pursued the arts instead of practicing medicine. He is known for his paintings of large bodies of water in which the landscape is reflected in soft, diffused light. However, the most famous painting of the Hudson River School was a work by Asher B. Durand, a dramatic four foot by three foot painting titled *Kindred Spirits.* In it, Durand stands beside Thomas Cole, tiny figures on a rocky promontory in the Catskills. In the background are the Catskill Falls. In the foreground is a broken tree stump, which symbolizes the temporary condition of life. The mood expresses the harmony between man and nature.

3. According to paragraph 5, which artist painted the work that is best known as representative of the Hudson River School?
   - Ⓐ Cole
   - Ⓑ Church
   - Ⓒ Gilford
   - Ⓓ Durand

Paragraph 5 is marked with an arrow ➔.

4. Which of the sentences below best expresses the information in the highlighted sentence in the passage? The other choices change the meaning or leave out important information.
   - Ⓐ The contemporary members of the Hudson River School had forgotten about Church.
   - Ⓑ The Hudson River School was forgotten after the passing of Church and the original members.
   - Ⓒ All of the members of the Hudson River School except Church were forgotten when they passed away.
   - Ⓓ Church did not remember the contemporary artists of the Hudson River School when they passed away.

5. According to the passage, everything is true about the Hudson River School EXCEPT
   - Ⓐ Members often traveled abroad to explore new landscapes.
   - Ⓑ They always painted their works on site in the open air.
   - Ⓒ Their paintings inspired the settlement of the Western U.S.
   - Ⓓ Some younger members studied at the Dusseldorf Academy.

By the time that Church and his contemporaries had passed away, the Hudson River School had been all but forgotten. After the Civil War, the Barbizon School had virtually replaced it. In addition, portrait painting had become more popular, attracting at least some of the noteworthy artists at the expense of landscape themes. Nevertheless, in addition to the body of work that was produced by the members of the Hudson River School, it is fair to say that they contributed to the interest in the American wilderness that influenced both Western expansion and the preservation of many natural wonders that were discovered by explorers and settlers.

Reading Strategy 19

Make an intelligent guess

Never leave a question blank. To make an intelligent guess, eliminate as many answer choices as possible. The more choices that you eliminate, the better your chances are to choose the correct answer. If you decide to change your guess because of information that you learn later in the test, change your first answer, but if you are not absolutely sure, leave your original guess as your first and final answer. Many studies indicate that your first guess is more likely to be correct than a change that you make later.

EXAMPLE QUESTION

Why do scientists hypothesize that life may have existed on Venus?

Ⓐ Probes have located microbes on the planet's surface.

✗ The oceans have evidence of early life forms.

Ⓒ Water molecules have been found in the planet's atmosphere.

✗ Carbon dioxide suggests that plants have decomposed.

TIMED PASSAGE

Venus

Today Venus is too hot to support life in any of the forms known to the scientific community. It has an atmosphere primarily composed of carbon dioxide and sulfuric acid and a crushing surface pressure similar to that of the ocean floor on Earth. However, this inhospitable environment may have changed a great deal since the early history of the planet. Several probes have detected the evaporation of hydrogen and oxygen from Venus into space in proportions that suggest the historic existence of liquid water, either in the atmosphere or on the planet itself.

About four billion years ago, the Sun was only about 60 percent as bright as it is today, which means that both Earth and Mars would have been colder, and Venus may have been warm enough to support the evolution of warm oceans and the life forms that develop in them. Furthermore, cloud cover could have protected the oceans for several billion years, which means that for half of its existence, Venus could have supported life.

It is even possible, according to some scientists, that life on Venus was the origin of life of Earth. In the very early formation of the solar system, fragments of the forming planets were blasting off of the surfaces and, along with comets and meteors, were bombarding neighboring planets. If life was present on Venus, microbes could have traveled in the rocky bodies that bombarded Earth during its early stages of development. . . .

Practice 19

Choose the correct answer to the question. If you are sure that an answer is NOT correct, cross it off. Then if you have more than one possible answer remaining, make a guess. If you are not absolutely sure about an alternative answer after you have guessed, leave your first choice as your final answer. Check your answer in the Answer Key on page 313.

QUESTION

What caused the fur trade to diminish in importance?

Ⓐ Relationships between Europeans and Native Americans deteriorated.

Ⓑ The war in Europe interrupted trade with the Americans.

Ⓒ Clothing styles in Europe changed the demand for animal furs.

Ⓓ Trappers took jobs in cities after the Lewis and Clark expedition.

Fur Trade

The fur trade was one of the first and most important industries in North America, beginning in the 1500s with exchanges between Europeans and native peoples. Trappers traded their furs for tools, weapons, and other goods. The earliest fur traders were the French explorers who traveled throughout Eastern Canada. Initially, the exchange consisted of gifts to establish friendly relationships between the Europeans and the Native Americans, but eventually trading posts were established along the St. Lawrence River and the Great Lakes, and south along the Mississippi River. By 1670, the British Crown had given a group of English merchants exclusive trading rights to the Hudson Bay area. They established the Hudson Bay Company, which was the first of several large firms in the region. During the 1700s, the French and the British continued a bitter rivalry for trading rights, which eventually led to the French and Indian War, a conflict that ended in 1763 when Great Britain confiscated the French Colonial holdings in North America.

With the Lewis and Clark expedition in 1805, fur trading expanded into the West; however, since the Western tribes had less interest in trapping than the Eastern tribes, European and American mountain men sold their furs to the companies that were using pelts for clothing, especially beaver hats that were the rage throughout Europe. By the 1800s the fur industry had started to decline, in part because the animals had diminished as a result of overtrapping, and in part, because the European hat manufacturers had begun to use silk instead of felt. In addition, the settlements that had grown up around the trading posts and forts were already growing into major cities such as Montreal and Quebec in Canada and Detroit, St. Louis, and New Orleans in the United States. It was, in fact, the claims made by fur traders that determined the border between the United States and Canada. Although the initial exchange of goods for furs had established positive relationships with the tribes in the fur-trading region, the settlements that encroached on their traditional land caused hostility between the Europeans and the Native Americans.

★★★**Bonus:** iBT Listening, ITP Listening, Structure
You can use this strategy to answer questions in sections of the TOEFL where you must choose among four answer choices.

Reading Strategy 20

Use the review screen efficiently

Click on the **Review** button at the end of each passage to make sure that you have not left any questions blank in that passage. If you have not attempted a question, you can find it on the review chart, click, and return to it in the passage. In general, you should try to answer all of the questions as you move through the test. If you use the **Review** button too often, you may not have enough time to finish the last passage.

EXAMPLE SCREENS

REVIEW SCREEN

The following screen is from the CD-ROM that accompanies *Barron's TOEFL iBT*. It is very similar to the Review screen on the actual exam. On this Practice Test, Questions 27, 34, and 37 were not attempted. The review was accessed at the end of the passage on Geothermal Energy. Although it is not a good idea to leave questions blank while you are moving through the test, the Review screen helps you find questions that you did not answer.

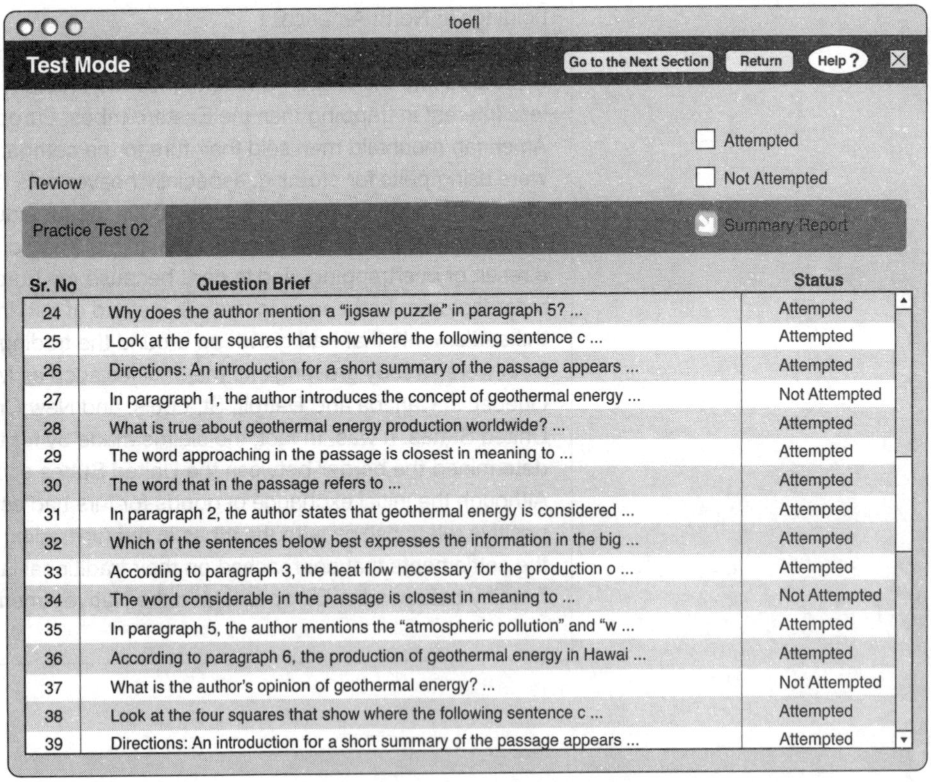

Sr. No	Question Brief	Status
24	Why does the author mention a "jigsaw puzzle" in paragraph 5? ...	Attempted
25	Look at the four squares that show where the following sentence c ...	Attempted
26	Directions: An introduction for a short summary of the passage appears ...	Attempted
27	In paragraph 1, the author introduces the concept of geothermal energy ...	Not Attempted
28	What is true about geothermal energy production worldwide? ...	Attempted
29	The word approaching in the passage is closest in meaning to ...	Attempted
30	The word that in the passage refers to ...	Attempted
31	In paragraph 2, the author states that geothermal energy is considered ...	Attempted
32	Which of the sentences below best expresses the information in the big ...	Attempted
33	According to paragraph 3, the heat flow necessary for the production o ...	Attempted
34	The word considerable in the passage is closest in meaning to ...	Not Attempted
35	In paragraph 5, the author mentions the "atmospheric pollution" and "w ...	Attempted
36	According to paragraph 6, the production of geothermal energy in Hawai ...	Attempted
37	What is the author's opinion of geothermal energy? ...	Not Attempted
38	Look at the four squares that show where the following sentence c ...	Attempted
39	Directions: An introduction for a short summary of the passage appears ...	Attempted

Practice 20

Answer the following questions to practice your skills on the Reading screens. Check your answers with those in the Answer Key on page 313.

1. What happens when you click on the **Review** button?

2. What is the purpose of the **Review** screen?

3. When should you click on the **Review** button?

4. How often should you use the **Review** button?

5. Why would you leave a question blank on the Reading section?

3

STRATEGIES FOR THE LISTENING SECTION

OVERVIEW OF THE iBT LISTENING SECTION

The Listening section is Section 2 on the Internet-Based TOEFL. The Listening section tests your ability to understand spoken English that is typical of interactions and academic speech on college campuses. During the test, you will listen to conversations, lectures, and discussions, and you will answer questions about them.

There are two formats for the Listening section. On the short format, you will listen to two conversations, two lectures, and two discussions. On the long format, you will listen to three conversations, three lectures, and three discussions. After each listening passage, you will answer five or six questions about it. Only two conversations, two lectures, and two discussions will be graded. The other passages are part of an experimental section for future tests. Because you will not know which conversations, lectures, and discussions will be graded, you must try to do your best on all of them.

You will hear each passage one time. You may take notes while you listen, but notes are not graded. You may use your notes to answer the questions.

Choose the best answer for multiple-choice questions. Follow the directions on the screen for computer-assisted questions. Click on **Next** and then on **OK** to go on to the next question. You cannot return to previous questions.

The Listening section is divided into sets. Each set includes one conversation, two lectures, and two discussions. You will have 20 minutes to answer all of the questions on the short format and 30 minutes to answer all of the questions on the long format. A clock on the screen will show you how much time you have to complete your answers for the section. The clock does NOT count the time you are listening to the conversations, lectures, and discussions.

OVERVIEW OF THE ITP LISTENING COMPREHENSION SECTION

The Listening Comprehension section is Section 1 on the Paper-Based ITP. The Listening Comprehension section tests your ability to understand spoken English. During the test, you will listen to dialogues, longer conversations, and talks, and answer questions about them.

In Part A, you will hear short dialogues between two people. After each exchange, you will hear a question about the information in the dialogue. In Part B, you will hear longer conversations. After each conversation, you will hear several questions. In Part C, you will hear several short talks. After each talk, you will hear several questions.

You will hear each dialogue, conversation, or talk one time. The questions are not printed in your test book. You may NOT take notes while you listen and you may NOT write in your test book.

Choose the best answer from four possible answers written in your test book. Then, on your answer sheet, fill in the oval that corresponds to the letter of the answer you have chosen.

You will have 35 minutes to listen and answer all of the questions.

TIPS FROM SUCCESSFUL TOEFL STUDENTS

WHAT TO DO ON THE LISTENING SECTION

- Look for opportunities to attend free lectures in English. If you are near a university that offers academic courses in English, ask the professors if you can listen to a lecture or two in their classes.
- If you have access to television in English, watch the Discovery Channel, National Geographic Channel, Smithsonian Channel, and other stations that offer academic topics. Don't pay too much attention to the screen. Take notes while you listen.
- When you are taking notes, don't try to write down every word. You are not required to take a dictation. Just jot down what you need to remember the main ideas and important details.
- Get familiar with the directions. If you don't have to read the directions and think about what to do, you will save time. Be sure that you understand how to use the **OK** button. On the Listening section, you have to click on **Next** and then on **OK** to go to the next question. If you forget to click on **OK**, you will not move to the next screen.
- Don't click on the **Help** button unless you absolutely don't understand what to do. The only information in the **Help** menu is a recap of the directions. And the clock keeps running!

Listening Strategy 1

Improve your concentration

Focus on the speaker and on your notes and learn to bring your attention back from external disturbances and internal distractions. External disturbances may include someone coughing or making noise. Sometimes others will begin the Speaking section while you are finishing the Listening section or you may see movement out of the corner of your eye. To control your response to these distractions, focus your eyes on the notes you are taking. Don't follow the noise or movement with your eyes because your mind will also move away from the task. Internal distractions are emotions that cause you to think about something else. If you don't like the speaker's voice, if you can't understand a word, or if you don't know the answer to a question, you cannot allow yourself to get upset and start a worried conversation in your mind. This will cause you to miss more information in the passages. Learn to say "No" when you start to panic. Take a deep breath and focus on the passage or the current question instead of worrying about the last question.

EXAMPLE DISTRACTION	EXAMPLE RESPONSE
The speaker's voice is irritating.	Focus on the speaker's words, not the voice.
Another student is coughing.	Do not look up. Stay focused on the notes.
You don't know the answer to a question.	Guess and move on to the next question.
You feel afraid that you are not doing well.	Take a breath. Focus on the notes.
Someone is leaving the room.	Do not look at the person. Focus on the notes.

Practice 1

First, listen for the orientation and identify the subject. Then, listen for more information and use your focus skills when distractions occur. Take notes and check them using the example notes in the Answer Key on page 314.

CD 1, Track 2

Notes

★★**Bonus:** iBT Speaking, Writing
Strategies that help you concentrate on the Listening section will also help you on the parts of the iBT Speaking and Writing sections that include listening passages.

Listening Strategy 2

Ignore the photographs of people in an academic setting

You will see photographs of people in a classroom at the beginning of each lecture and discussion and you will see photographs of people in offices and on campus at the beginning of each conversation. Be careful. These photographs do not add information to the listening passages. Don't get distracted. Pay attention to the audio and don't focus on the people's faces, their clothing, the classroom, the office, or the objects in the setting. You will miss hearing the information that you need to answer the questions.

LISTENING LECTURE

EXAMPLE PHOTO

If you allow your mind to focus on the photograph, you will probably begin looking at the teacher, her hair, her clothing, her jewelry, and the expression on her face. Your focus may then stray to the blackboard, and you may try to read the information written on it. Then you may look at the students. You may count how many are in the photograph, whether they are men or women, and you may start to think about the color of their clothing, their hair, their ethnicity, or other aspects of the classroom setting. None of this visual information will be included in the questions about the lecture.

Practice 2

Listen to the lecture and take notes. Do not allow your mind to focus on the photograph of the classroom setting. Compare your notes with the example notes in the Answer Key on page 315.

CD 1, Track 3

Notes

Listening Strategy 3

Listen for ideas

You do not need to recognize each individual word in order to understand the general ideas in a conversation or lecture. In fact, if you stop to think about a new word, you will probably miss the next word or even the next sentence. If that happens, you can get confused. You will probably not be able to recognize all of the words, but you will be able to understand the ideas if you don't lose your concentration when the speaker uses an unfamiliar word. In the example script, some words are underlined. Even if you do not understand them, you can get the main idea of the listening passage.

EXAMPLE SCRIPT

Navajo Textiles

The Santa Fe Railroad built a line from the eastern United States to the Four Corners Area in the 1890s, making it possible for tourists to travel to the Navajo Territories. Although they had been making textiles since at least 1700, a skill that they had learned from the Pueblo tribes, the Navajos had been weaving cloth for practical uses such as bed coverings, saddle blankets, and even cloaks for the tribal chiefs. The materials that they used for their textiles were all locally gathered. Although the Pueblo had used cotton, the Navajo preferred yarn spun from the wool of sheep that they raised. The dyes for the yarn were mixed from the colors found in local plants and the designs included a wide variety of geometric figures that represented kinship ties, regional affiliations, or spiritual symbols.

When European tourists were introduced to the textiles, they appreciated them for their artistic value as well as for their practical use. They compared them with the Oriental floor coverings that were popular at the time and made purchases with a view to using them as rugs. The huge demand for the rugs brought about several changes in the manufacture. First, Navajo rug makers began to rely on commercial wool yarn instead of spinning their own. In addition, they experimented with new dyes and colors that were more attractive to the European buyers. Moreover, they introduced new designs, more like the patterns found in the Oriental rugs that the tourists had compared with the Navajo textiles.

Trading Posts mushroomed along the railroad lines to facilitate the purchase of souvenirs, including Navajo rugs. At first, in an effort to meet the high demand in the shortest possible time, some of the rugs were not as well made as the traditional textiles had been; however, artisans soon realized that it was more important to take the time to create the beautifully woven pieces that they had historically produced, selling them for a higher price point to tourists who were willing to pay for a uniquely crafted rug. . . .

Practice 3

Listen for general comprehension and take notes on the topic and the three major points. Some of the words are beeped so that you will not hear all of the individual words. Check your answers in the Answer Key on page 316.

CD 1, Track 4

Notes

Topic

Major Point 1

Major Point 2

Major Point 3

★★**Bonus:** iBT Speaking, Writing
In addition to the Listening sections on the ITP and the iBT, getting the gist of a listening passage will be important on the Speaking and Writing sections of the iBT which include listening passages.

Listening Strategy 4

Expect to hear natural speech

You will be listening to natural speech in the listening passages on the iBT. Professors and students will interrupt themselves, hesitate, repeat, and say *uh* or *um*, *well* or *you know* to give themselves time to think. All of this is very different from the clear speech that you have probably heard on traditional English tests. In order to succeed on the Listening section, you need to have experience listening to natural speech.

Example Script

Red Tides

. . . so when you see that red color in marine coastal areas, uh, when you see that, you could be seeing the red tide that we have been talking about. But remember, the more correct term is harmful algal bloom or HAB because, uh, the occurrence, of the red tide, I mean, it really doesn't have anything to do with the tides, and of course, not all algal blooms cause that red discoloration of the water. Some are green or brown.

Okay then, we know that typically, HABs occur when algae—single-celled organisms such as, for example, uh, phytoplankton—when they reproduce rapidly. In low concentrations, these organisms are harmless, but in higher concentrations, well, uh . . . the toxins that they release . . . these toxins . . .paralyze the..the central nervous systems of fish in the area and consequently, the fish die and wash up on the shore.

But what causes the increase in the algae in the first place? Well, HABs are . . . HABS often result from an increase in nutrients that algae need, especially nitrates and phosphates. And these chemicals—the nitrates and phosphates—we need to think about how they get into the coastal areas. So let's discuss three ways that it might happen. First, let's consider, uh, runoff from farming operations. Pesticides, irrigation water, fertilizer—all of these can, uh, they cause an imbalance in the nitrate and phosphate levels, and they get dumped into the water from agricultural areas. Then there is water pollution from manufacturing which . . . I'm not saying that it always contains nitrates and phosphates, but it can. For example, when soaps and detergents are disposed of in the water system, they contain high levels of phosphates. But sometimes natural events such as a . . . an increase in water temperature can also affect the concentration of algae because it grows faster in a warm environment. Also changes in the salt content of the water or even a prolonged period of very calm seas can, uh, can cause an HAB cycle to begin . . .

Examples of Natural Speech

Interrupts himself
Restarts the sentence
Hesitates
Uses verbal thought pauses—*uh* and *um*
Repeats phrases

Practice 4

Listen to the lecture and take notes. Do not allow your mind to focus on the hesitations and verbal thought pauses. Compare your notes with the example notes in the Answer Key on page 317.

CD 1, Track 5

Notes

★★**Bonus:** iBT Speaking, Writing
Learning to listen to natural speech will be helpful not only on the iBT in the sections of the test that include listening passages, but also in the real world when you are having a conversation.

Listening Strategy 5

Differentiate between the main idea of a lecture and classroom business

You will usually hear the main idea of a lecture in the narrator's introduction or in the first few sentences, but sometimes the professor will conduct classroom business before beginning the main topic. Examples of classroom business include information about exams, tests, and quizzes, changes in the class schedule, references to the syllabus, classroom policies, and clarification of assignments. You may be asked details about classroom business, but even though you hear this introductory information first, the main idea follows. Transition words and phrases that often signal the introduction of the main idea are listed below.

Signal Words: Classroom Business	***Transition Words: Lecture***
Before we begin, let me remind you . . .	*Okay . . . .*
Sorry that I don't have the exams graded . . .	*Okay then . . .*
Good job on the quizzes . . .	*So . . .*
Let's take a look at the syllabus . . .	*Now . . .*
Let me remind you that . . .	*Good . . .*

Example Script

Attitude and the Immune System

<u>There seems to be some confusion about my attendance policy, so before we begin today, I want to refer you to the syllabus where I have it spelled out.</u> It's on page 2, and uh, as you can see, you have one excused absence without explanation, but if you must miss class again, then you need to get in touch with me, and I have to know what the problem is so that we can work it out. If you miss three classes, and uh, that's for whatever reason, then . . . then your grade will be lowered by one letter. <u>Okay then</u>, I hope that clarifies any misunderstanding. <u>Now, let's turn our attention to the immune system, and specifically, to the relationship between attitude and a healthy immune system</u>. Uh, it has been generally recognized for a long time that a positive attitude contributes to general well being, and stress causes at least a subjective feeling of ill health, but those previous studies, they have compared optimists with pessimists with generalities that were interesting, but didn't really answer the question of how the immune system responds in a single person. So, I am talking about immunity in the cell structure, and that is something quite different.

New studies published in reputable journals like *Psychological Science* suggest that when optimism increases in a subject, the immune system also improves; conversely, when a subject begins to engage in pessimistic thoughts and feelings, the immune system responds. And it's important to mention here that the immune system is now recognized as being closely integrated with the nervous system and other bodily functions. To put that a different way, the immune system is very important to many other systems in the body, and for optimal health, the immune system must be functioning at optimal levels.

With modern methods, we can actually track the cellular efficiency of the immune system, we can . . . we can compare the rise and fall of optimism in a subject, and confirm the fact that a positive attitude at any one time in an individual can significantly affect the functioning of that person's immune system . . .

Practice 5

First, listen for the signal words and identify the topic of the classroom business. Then, listen for the transition words and identify the topic of this lecture. What is the main idea? Check your answers in the Answer Key on page 318.

CD 1, Track 6

Notes

Signal words: ______________________________

Topic for classroom business: ______________________________

Transition words: ______________________________

Topic for this lecture: ______________________________

Listening Strategy 6

Distinguish between the main idea of a lecture and a review

Professors typically begin a lecture or discussion by introducing the main idea, but sometimes they will refer to the topic of the previous lecture before introducing the topic of the current lecture. Occasionally, the lecturer will also present a brief review of the previous lecture in order to provide background for new information. Be sure to listen carefully to hear the references to the times for each topic. Be careful not to confuse the main idea of the previous lecture with the main idea of the current lecture. Some typical phrases that signal times for topics include those listed below.

Signal Words: Previous Lecture	***Transition Words: Current Lecture***
As you will recall . . .	*Today we are going to . . .*
You will remember that . . .	*Today we will discuss . . .*
In the last class . . .	*Now let's turn our attention to . . .*
In the previous lecture . . .	*Today's lecture will . . .*
We were discussing . . .	*With that in mind, let's talk about . . .*

EXAMPLE SCRIPT

The Trickster as a Universal Character

As you will recall from last week's lecture, the oral tradition is found in all cultures in spite of the relatively recent reliance on written traditions by literate societies. To review a little bit, oral traditions are important to the archiving of historical records and the education of the younger generations as well as to the preservation of religious practices. We talked about the fact that oral traditions often require that the persons relating the information repeat it without altering the record. Of course, variations may occur, but the oral account may be corrected or criticized in the same way that a written account is subject to, uh, review and criticism. I think I also mentioned that some of the stories or other information can be related only at certain times, within specific contexts by elders or members of society who are considered worthy of being entrusted with special knowledge, and occasionally, a gift is expected for a story to be told or . . . or a religious ceremony to be performed.

Okay. That brings us to today's topic. Although it's true that stories are specific to the cultures that they represent, a relatively small number of themes and characters figure in the oral traditions of . . . well, virtually all world cultures. An example of a universal character that comes to mind, uh, that I want to talk about. . . .The example is the trickster, and I want to spend some time discussing it because the trickster is such an iconic character . . . a comic character who breaks the rules of society, often by playing practical jokes or by petty theft. So the role of the trickster is to play tricks and engage in games or mischief that's usually viewed as humorous rather than malicious. Coyote and Raven are obvious examples of tricksters that appear in the oral traditions of Native American tribes in North America. The classical Greek and Roman myths include tricksters like Prometheus who steals fire from the gods. And, uh, how about the Monkey King in Chinese traditions? I think he also qualifies as a trickster. So, we see that many examples of tricksters are found around the world . . .

Practice 6

First, listen for the signal words and identify the topic of the previous lecture. Then, listen for the transition words and identify the topic of this lecture. What is the main idea? Check your answers in the Answer Key on page 319.

CD 1, Track 7

Notes

Signal words: ____________________

Topic for previous lecture: ____________________

Transition words: ____________________

Topic for this lecture: ____________________

Listening Strategy 7

Listen for the purpose of a conversation after a greeting

When you hear a conference in a professor's office, the speakers usually state the purpose of the conference after a short social exchange. Either the professor asks the student the reason for the appointment or the student explains why he or she wants to speak with the professor. When you hear the end of the greeting, listen carefully for the purpose of the conference. The first question for a conference passage usually asks you to choose the answer choice that identifies the purpose of the conversation.

Typical Greetings

How's it going?
Hi, Marge.
How are you doing?
Excuse me, Professor Lee.
Hey, Greg.

Direct Questions

What can I do for you?
What do you have on your mind?
How can I help you?
Come in, Ted. What's going on?
What do you want to talk about?
What's up?
What seems to be the problem?

EXAMPLE SCRIPT

Greeting

Professor: Hi Susan. How's it going?

Susan: Fine. Thanks for seeing me.

Professor: No problem. What can I do for you?

Purpose

Susan: Well, I'm trying to make a decision about my courses next semester. And today's the last day of registration.

Professor: Okay . . .

Susan: You see, I have to take one more foreign language class—another Spanish class—to fulfill the requirement. But they're five-hour classes so they meet every day, and then there's the lab. So I'd, uh, I'd really like to take a break, from the Spanish classes, I mean . . . but maybe that's not a good idea.

Professor: Because?

Susan: Because I might forget some of the Spanish I've already learned, and it might be harder to get back into it after the break.

Professor: Oh, I see. And why do you want to take a break?

Susan: I live in Springfield, and that's about forty miles from campus, so driving in every day is . . . is . . .

Professor: . . . taking a lot of time?

Susan: Exactly. Plus, it's winter quarter, and that means the roads are going to be bad, at least some of the time. Last winter I missed a lot of days because I couldn't get to class.

Professor: Right. Well, have you considered taking an online course? You could do that from home.

Susan: I thought about that, but then I was thinking that it wouldn't give me much of an opportunity for conversation, in Spanish, uh, with the other students.

Professor: That could be a disadvantage, but I think there are a lot of listening options in the online course curriculum.

Susan: Oh. Well, uh, do you know who teaches the online course, for 104?

Professor: As a matter of fact, it's Mr. Martin.

Susan: And he's a native speaker, isn't he? I like to take my language classes from native speakers.

Professor: I'm almost sure that he is. But, listen, why don't you talk with him about the online courses? His office is just across the hall, and I think I saw him come in a few minutes ago . . .

Practice 7

First, listen for the greeting and then be ready to identify the purpose of the conference. Write down the information that you hear after the greeting. Check your answers in the Answer Key on page 320.

CD 1, Track 8

Notes

1. **Purpose of conference:** ______________________________

2. **Purpose of conference:** ______________________________

3. **Purpose of conference:** ______________________________

4. **Purpose of conference:** ______________________________

5. **Purpose of conference:** ______________________________

★**Bonus:** iBT Speaking
You can also use this strategy to identify the purpose of conversations on Tasks 3 and 5 in the Speaking section of the iBT.

Listening Strategy 8

Pay attention to blackboards

When you see a blackboard with a word or phrase written on it, copy the word or phrase in your notes and listen carefully when the lecturer explains the term or concept. A blackboard is a visual signal that the information is important.

Example Blackboard

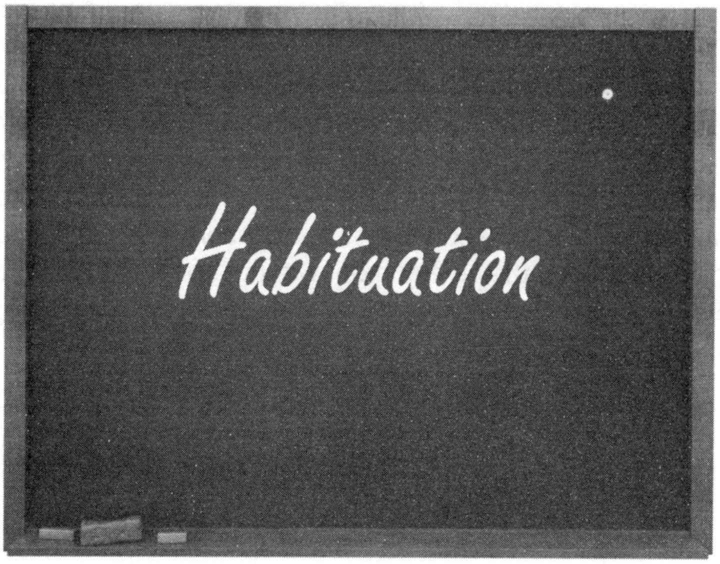

EXAMPLE SCRIPT

The human brain is programmed to respond to something that is new or different. But, after repetition, for example, when you have the same experience repeatedly, then the brain stops responding at the same level of attention. Think about it. When an event is familiar you may start responding slower or, or . . . you may even stop responding altogether. For example, if you move into a new neighborhood and you hear a noise for the first time, you will direct your attention to it, like traffic noise from a freeway nearby. But if the same noise, the traffic noise, if it happens again and again, you probably won't pay much attention to it. Or, if you smell something in the work environment, you will notice it initially, but then, after some time, you will notice it less or not at all. In psychology, we call this *habituation*. So *habituation* is a decline in responsiveness to a repeated stimulus or situation. And this is true not only for humans but also in the animal world.

Let me give you an example of habituation in birds. In a classic study by Tinbergen, I think it was about 1950, birds responded strongly the first time that the investigators passed the shadow of a hawk overhead, but when the shadow continued to appear with no attack, then the birds began to ignore it. In psychology, we might put it this way: The initial stimulus elicited a startle response, but after repetition, the birds became habituated to the experience. And this is logical when we think about it because in a very real sense, that initial orientation to novelty is a survival mechanism. But through repeated experiences, humans and animals learn to identify and ignore events that occur regularly without serious benefits or consequences. Habituation is useful because it allows us to filter large amounts of information in the environment without spending time investigating every sensory experience every time we have it, something that would be virtually impossible to do. Arguably habituation is one of the most basic and important forms of learning and . . .

Practice 8

When a blackboard appears, copy the word or phrase that you see on the blackboard and listen for a definition or explanation. Compare your notes with the example notes in the Answer Key on page 321.

CD 1, Track 9

Notes

★★**Bonus:** iBT Reading, ITP Reading
The ITP Listening does not include blackboards but the definition of terms is common in the ITP Listening section. When you learn to use blackboards, you also learn to identify terms, which is a good strategy for the Reading sections.

Listening Strategy 9

Watch for visual cues in the answer choices

You will see ovals or boxes beside the answer choices for multiple-choice questions. When you see *ovals*, you know that you will be looking for one answer. When you see *boxes*, you know that you will be looking for two or more answers. If you choose one answer for a two-answer question, a warning screen will appear, and you will waste time getting back to the question screen. Every time you see squares, make the correct number of answer choices before moving to the next question.

Example Questions

1. Which of the following works is attributed to Homer?

   ○ *The Journey*

   ○ *The Voyage*

   ● *The Odyssey*

   ○ *The Conquest*

2. Which of the following works are attributed to Homer?

   Click on 2 answer choices.

   ☐ *The Journey*

   ☐ *The Voyage*

   ☒ *The Odyssey*

   ☒ *The Iliad*

Practice 9

Look for the visual cues and identify the number of answer choices to mark for each question. You are not expected to know which answers, just the number of answers to select. Then check your answers in the Answer Key on page 322.

CD 1, Track 10

1. Which of the following marine animals are Cetaceans?
   - ☐ Dolphins
   - ☐ Sharks
   - ☐ Whales
   - ☐ Seals

2. According to the professor, which of the following days of the week were named for the Norse gods?
   - ☐ Tuesday
   - ☐ Thursday
   - ☐ Saturday
   - ☐ Sunday

3. According to the passage, what do we know about oil paint?
   - ○ It does not allow the artist to blend and refine colors.
   - ○ It tends to fade to a lighter color with age.
   - ○ It is possible to paint over mistakes on the canvas.
   - ○ It dries evenly and quickly after application.

4. Which of the following are characteristic of viruses?
   - ☐ They are much smaller than bacteria.
   - ☐ They must grow within a host.
   - ☐ They have a nucleus in their cell structure.
   - ☐ They respond to antibiotics.

5. According to the passage, which Canadian provinces are included in the Maritimes?
   - ○ Newfoundland
   - ○ Quebec
   - ○ Nova Scotia
   - ○ Yukon

Listening Strategy 10

Listen for verbal signals to pay attention

When you hear verbal signals in a lecture or discussion, listen carefully and take notes. You will probably find the information that you hear after the signal in one of the test questions. Examples of the most common verbal signals are listed below.

Signal Words: Attention
Now this is important...
The point is . . .
Remember that . . .
The basic concept is . . .
The issue is . . .

Example Script

Interchangeable Parts

. . . so we see that before the late 1700s, interchangeable parts were not part of the manufacturing process. Every item—guns, clocks, wagons—you name it . . . every item was crafted by hand with parts that were made specifically for that item. It was not only an expensive process, but . . . and this is important, it was not a very reliable process. In the case of guns, for example, they commonly misfired and had to be sent away for repair, which was inconvenient on the farm but disastrous on the battlefield. So several gun manufacturers came up with the idea of making guns with interchangeable parts. In fact, one of the inventors, Eli Whitney, who had previously invented the cotton gin . . . he made ten guns with standardized parts that . . . that were interchangeable. And he took the guns to a session of Congress to demonstrate how the process would work. First he disassembled all of the guns and mixed up the parts. Then he reassembled them and he proved that all of the parts could be used interchangeably in the guns. So when a soldier's gun needed repair, he could put in a new part that would fit perfectly instead of sending the gun to a craftsman for a lengthy repair.

Okay, today the process seems well, just logical, but at the time, it was a revolutionary idea to manufacture standard replacement parts. The point is that interchangeable parts set the stage for the Industrial Revolution because it was a short step from the idea that parts could be standardized to the idea that items could be mass produced on an assembly line. Remember that the parts for Whitney's guns still had to be made by a craftsman, but to rigid specifications, of course, and the craftsman made all of the individual parts for the gun. In an assembly line, workers specialized in making just one part which allowed them to use a limited number of tools and to work more efficiently. And the basic concept here is that the items could be mass produced at a lower price and made more available to the public . . .

Practice 10

First, listen for the signal words and identify the important information. Write down the information that you hear after the signal words. Check your answers in the Answer Key on page 323.

CD 1, Track 11

Notes

1. Signal words: ______________________________

   Important information: ______________________________

2. Signal words: ______________________________

   Important information: ______________________________

3. Signal words: ______________________________

   Important information: ______________________________

4. Signal words: ______________________________

   Important information: ______________________________

5. Signal words: ______________________________

   Important information: ______________________________

★★**Bonus:** iBT Speaking, Writing
In addition to the lectures on the ITP and iBT Listening section, you can also use this strategy to identify important information in short lectures in Tasks 4 and 6 of the iBT Speaking section and the lecture in the Integrated Essay of the iBT Writing section.

Listening Strategy 11

Take careful notes when you hear repetition and restatement

When you hear repetition and restatement in a lecture or discussion, you know that the information is important and you should include it in your notes. *Repetition* is saying the same thing in the same way. *Restatement* is saying the same thing in a different way. Sometimes the repetition or restatement is introduced with a signal phrase like those listed below.

Signal Words: Repetition	***Transition Words: Restatement***
Let me say that again . . .	*I mean . . .*
Let me repeat that . . .	*To say that another way . . .*
	In other words . . .

Example Script

Olber's Paradox

From the answers on the last quiz, I see that some of you are still having problems with Olber's paradox, so I'm going to explain it in a different way today. First, let's review the question: Why is the sky dark at night? If the universe is static and eternal, and if it is infinite and uniformly filled with stars, then we should be able to look in any direction in the sky, and we should be looking at a star, regardless of how far away it might be. Let me repeat that: If the universe is static and eternal, and if it is infinite and uniformly filled with stars, then we should be able to look in any direction in the sky, and we should be looking at a star, regardless of how far away it might be. Okay, now if that is so, then the line of sight from the Earth will end at the surface of a star. So the sky at night should be bright, not dark. Right? But clearly, this contradicts the fact that the night sky is dark.

Well, think about this. In order to explain Olber's paradox, we have to look at the assumptions in the paradox. In other words, the assumptions are incorrect. So let's try to break down the assumptions. In the first place, galaxies are not static. They are moving away from us, which explains why the energy cannot be detected. Second, the universe is not eternal since it was created in a moment. Remember one of the theories for the origin of the universe is referred to as the Big Bang. To say that another way, if the universe was created at some specific point in time, maybe light from the most distant stars has not had enough time to reach the Earth. Or, perhaps the universe is eternal, but stars have only existed for part of that time, so then only starlight from a certain distance would reach the Earth. By that I mean that while the universe may contain an infinite number of stars, we can probably only see those that lie within our limited horizon. So you see . . .

Practice 11

Listen for signal words that identify repetition or restatement. Write down the information that you hear after the signal words. Check your answers in the Answer Key on page 324.

CD 1, Track 12

Notes

1. Signal words: ____________________

   Restatement: ____________________

2. Signal words: ____________________

   Repetition: ____________________

3. Signal words: ____________________

   Restatement: ____________________

4. Signal words: ____________________

   Restatement: ____________________

5. Signal words: ____________________

   Repetition: ____________________

★★**Bonus:** iBT Speaking, Writing
In addition to the Listening sections on the ITP and the iBT, you can also use this strategy to identify important information in short lectures in Tasks 4 and 6 of the iBT Speaking section and the lecture in the Integrated Essay of the iBT Writing section.

Listening Strategy 12

Identify the purpose of questions

You need to understand why a professor asks a question during a lecture or discussion. Sometimes professors ask real questions to encourage participation and student responses, but other times they ask rhetorical questions that they expect to answer themselves to clarify or expand on the information they are presenting. When professors ask rhetorical questions, they want you to think about the answers without offering verbal responses. When you hear a rhetorical question, it may also be a signal that the information in the answer is important.

Example Script

Cognitive Dissonance Theory

Cognitive dissonance theory was developed in the late 1950s by social psychologist Leon Festinger. Actually, it is a fairly simple concept, but it remains a very important one. First, let me explain *cognition*, which is just knowledge of a fact. I am a woman. This is a psychology class. Okay, so we all have a lot of cognitions, countless cognitions, in our heads at the same time, and most of them don't have any relationship to each other. But, some of them, well they *are* related in some way. Let's say, for example, that I want to be healthy, and I eat healthy snacks. That follows, right? <u>But what if I have two thoughts that don't go together? What if I want to be healthy, but I smoke?</u> With these conflicting ideas, I will probably start to feel uneasy. That's dissonance. So cognitive dissonance then, is holding inconsistent attitudes, beliefs, opinions, or values simultaneously, and that inconsistency—that causes inner conflict.

I have already alluded to a classic example of cognitive dissonance, that is, the smoker who believes that smoking is unhealthy. <u>Okay, so how can the smoker resolve the conflict? How can he deal with the cognitive dissonance?</u> Well first, the smoker may rationalize the behavior. For example, the smoker may deny the amount of smoking that he actually engages in or he may refute the research because it is done with laboratory animals, not humans. Or the smoker may deny that people who smoke die of lung cancer because he does not know anyone personally who smoked and who had terminal lung cancer. Probably the most common rationalization is that although smoking is unhealthy for most people, the smoker sees himself as an exception. Or, instead of rationalizing, the smoker may introduce another factor that could affect the outcome. For example, he may claim that he exercises more than the people who get lung cancer, or he takes vitamins or supplements that protect him in some way.

But, interestingly enough, the most common way that people deal with cognitive dissonance is to ignore it or avoid it. The smoker simply refuses to read studies on smoking or listen to information about it. So a person must recognize that the inconsistency exists in order to experience enough psychological pressure to be motivated to change either the attitude or the conflicting behavior . . .

Practice 12

First listen for three rhetorical questions. Then write down the answers to the questions. Compare your answers with those in the Answer Key on page 325.

CD 1, Track 13

Notes

1. Rhetorical question: ______________________________

   Answer: ______________________________

2. Rhetorical question: ______________________________

   Answer: ______________________________

3. Rhetorical question: ______________________________

   Answer: ______________________________

★**Bonus:** iBT Speaking
You can also use this strategy in the Speaking section. Ask and answer a rhetorical question when you are summarizing the lecture in Task 6.

Listening Strategy 13

Listen for intonation to express surprise

Sometimes the intonation that you hear will change the meaning of the words. A sentence with a falling intonation at the end is usually neutral. The same sentence with a rising intonation at the end can express surprise. Notice that the subject-verb order of the words does not change with the intonation.

EXAMPLE SCRIPT

CD 1, Track 14

Neutral: That's a good deal for a meal plan.
Surprised: That's a good deal for a meal plan.

Neutral: That's not a problem.
Surprised: That's not a problem.

Neutral: My advisor has to sign this form.
Surprised: My advisor has to sign this form.

Neutral: The door to the classroom is locked.
Surprised: The door to the classroom is locked.

Neutral: The room has been changed.
Surprised: The room has been changed.

Neutral: Tuition has gone up again.
Surprised: Tuition has gone up again.

Neutral: My name isn't spelled correctly on the transcript.
Surprised: My name isn't spelled correctly on the transcript.

Neutral: He let you borrow his notes.
Surprised: He let you borrow his notes.

Neutral: The time limit's up.
Surprised: The time limit's up.

Neutral: The test is today.
Surprised: The test is today.

CD 1, Track 14 PAUSE

Practice 13

First, listen to the sentence. Then identify the meaning as either neutral or surprised. Check your answers in the Answer Key on page 326.

CD 1, Track 14 RESUME

1. He's a student.	☐ Neutral	☐ Surprised
2. Susan is working full time at the library.	☐ Neutral	☐ Surprised
3. The bookstore is open.	☐ Neutral	☐ Surprised
4. I owe a late fee.	☐ Neutral	☐ Surprised
5. My advisor didn't approve it.	☐ Neutral	☐ Surprised
6. The class is off campus.	☐ Neutral	☐ Surprised
7. Dr. Wilson was there during his office hours.	☐ Neutral	☐ Surprised
8. You don't have my transcripts.	☐ Neutral	☐ Surprised
9. The T.A. was late again.	☐ Neutral	☐ Surprised
10. The class is cancelled.	☐ Neutral	☐ Surprised

★**Bonus:** iBT Speaking
In addition to conversations in both the iBT and the ITP Listening sections, you can use this strategy to recognize surprise in conversations in Tasks 3 and 5 of the Speaking section on the iBT.

Listening Strategy 14

Listen for intonation to express doubt

Remember that intonation can change the meaning of the words in a sentence. A sentence with a falling intonation at the end is usually neutral. The same sentence with a rising and then a falling intonation at the end can express doubt. It may help you to add "but I doubt it" in your mind after the doubtful sentence to repeat the intonation. For example, "That's what he said, but I doubt it."

Example Script

CD 1, Track 15

Neutral: That's what he said.
Doubtful: That's what he said.

Neutral: I think so.
Doubtful: I think so.

Neutral: That's okay.
Doubtful: That's okay.

Neutral: The quiz is usually easy.
Doubtful: The quiz is usually easy.

Neutral: I could go to the health center.
Doubtful: I could go to the health center.

Neutral: You can turn your paper in late.
Doubtful: You can turn your paper in late.

Neutral: Joe would give me a ride.
Doubtful: Joe would give me a ride.

Neutral: I can do that.
Doubtful: I can do that.

Neutral: You could live in the dorm.
Doubtful: You could live in the dorm.

Neutral: I have one more excused absence.
Doubtful: I have one more excused absence.

CD 1, Track 15 PAUSE

Practice 14

First, listen to the sentence. Then identify the meaning as either neutral or doubtful. Check your answers in the Answer Key on page 328.

CD 1, Track 15 RESUME

1.	Sometimes Professor Davis teaches the class online.	☐ Neutral	☐ Doubtful
2.	You could use your credit card.	☐ Neutral	☐ Doubtful
3.	The class might not be closed.	☐ Neutral	☐ Doubtful
4.	My roommate said she'd pay half of the bill.	☐ Neutral	☐ Doubtful
5.	You could ask for an excused absence.	☐ Neutral	☐ Doubtful
6.	Your book could be at the lost and found.	☐ Neutral	☐ Doubtful
7.	I can look on the Internet.	☐ Neutral	☐ Doubtful
8.	We can take a bus.	☐ Neutral	☐ Doubtful
9.	You could borrow Tom's notes.	☐ Neutral	☐ Doubtful
10.	We could take a break.	☐ Neutral	☐ Doubtful

★**Bonus:** iBT Speaking
In addition to conversations in both the iBT and the ITP Listening sections, you can use this strategy to recognize doubt in conversations in Tasks 3 and 5 of the Speaking section on the iBT.

Listening Strategy 15

Listen for tone of voice to express emotion

You will be asked to identify the speaker's emotions as well as the meaning of the words. Tone can be enthusiastic and interested; reserved, confused, apologetic, and unsure; or respectful, encouraging, and supportive. Listen to the enthusiasm in the tone of the professor in the first example. In the second example, the professor's tone is unsure, and in the third example, the professor is respectful and supportive of the participating students. Tone of voice can be very subtle sometimes. If you use TV to practice, don't look at the speaker. On the TOEFL, you won't have facial expressions to help you.

Example Script 1

CD 1, Track 16

Gullah

Professor: So although many people, even those who live on the coastlines of Georgia, South Carolina, and Florida, believe that Gullah is a Native American language, it is really a creole that was common among the black populations in the seventeenth century. It's actually a mix of several African languages and the English that was spoken in the 1600s when British explorers and settlers were beginning to populate the southeast coast of what would become the United States. More than 300 years later, it is remarkable that approximately 7,000 people speak Gullah as their only language and another 250,000 use Gullah with friends and family at least part of the time! Most speakers live on the sea islands off shore, but some mainlanders also speak Gullah. The relative isolation of the islands, which were accessible only by boat until the twentieth century, has probably contributed to the perpetuation of the language.

Since American colonists had little experience with the cultivation of rice, they were willing to pay higher prices for slaves from the traditional rice region of Africa who were used to laboring in the rice fields. Consequently, a large number of the slaves were from Senegal, Sierra Leone, and Liberia. Recent research has revealed enough similarities between them for Gullah and Krio, which is a creole language spoken in Sierra Leone, to be mutually comprehensible. However, most of the vocabulary for Gullah is English, whereas the grammar and phonetic systems are more like those of West African languages. Most linguists agree that its origin is a practical response to the problem of communication among slaves from different tribes and countries with no common language. In fact—and this is interesting—it's speculated that the word *Gullah* may have come from the name of the Angolan tribe that was identified as "N'gola" or "Gullahs." Still in question is whether the language derived from West African Pidgin English, which the slaves may have been familiar with before they were brought to the Americas, or whether it sprang to life on the rice plantations of Georgia, South Carolina, and Florida. So that's a mystery for us to unravel! We really don't know . . .

Example Script 2

Gullah

Professor: So uh, although many people, even those who live on the coastlines of Georgia, South Carolina, and Florida, they uh, they believe that Gullah is a Native American language, it's probably a creole that was common among the black populations in the seventeenth century. It could be a . . . a mix of several African languages and the English that was spoken in the 1600s when British explorers and settlers were beginning to populate the southeast coast of what would become the United States. More than 300 years later, approximately 7,000 people speak Gullah as their only language and another 250,000 use Gullah with friends and family at least part of the time. Most speakers live on the sea islands off shore, but some mainlanders also, we think that they speak Gullah. The relative isolation of the islands, which were accessible only by boat until the twentieth century, has probably, probably . . . contributed to the perpetuation of the language.

Since American colonists had little experience with the cultivation of rice, they were willing to pay higher prices for slaves from the traditional rice region of Africa who were used to laboring in the rice fields. Consequently, um . . . a large number of the slaves were from Senegal, Sierra Leone, and . . . and . . . Liberia. Recent research has revealed enough similarities between them for Gullah and Krio, which is a creole language spoken in Sierra Leone, for these languages to be mutually comprehensible. However, most of the vocabulary for Gullah is English, whereas the grammar and phonetic systems are more like those of West African languages. Well, most linguists agree that its origin is a practical response to the problem of communication among slaves from the different tribes and countries with no common language. In fact, it's speculated that the word *Gullah* may have come from the name of the Angolan tribe that was identified as "N'gola" or "Gullahs," but of course we can't be sure. Still in question is whether the language derived from West African Pidgin English, which the slaves may have been familiar with before they were brought to the Americas, or whether it sprang to life on the rice plantations of Georgia, South Carolina, and Florida. So, uh, we don't really know . . .

Example Script 3

Gullah

Professor: So although many people, even those who live on the coastlines of Georgia, South Carolina, and Florida, believe that Gullah is a Native American language, it is really a creole that was common among the black populations in the seventeenth century. It's actually a mix of several African languages and the English that was spoken in the 1600s when British explorers and settlers were beginning to populate the southeast coast of what would become the United States. More than 300 years later, it is remarkable that approximately 7,000 people speak Gullah as their only language and another 250,000 use Gullah with friends and family at least part of the time! So, what else have we learned about Gullah? Why is it still in use? Anyone?

Bill: Well, most speakers live on the sea islands off shore, but some mainlanders also speak Gullah, and the relative isolation of the islands, which were accessible only by boat until the twentieth century, has probably contributed to the perpetuation of the language.

Professor: Good observation, Bill. Since American colonists had little experience with the cultivation of rice, they were willing to pay higher prices for slaves from the traditional rice region of Africa who were used to laboring in the rice fields. Consequently, a large number of the slaves were from Senegal, Sierra Leone, and Liberia. So . . . Yes, Jane?

Jane: I just wanted to mention that recent research has revealed enough similarities between them for Gullah and Krio, which is a creole language spoken in Sierra Leone, for those languages to be mutually comprehensible. However, most of the vocabulary for Gullah is English, whereas the grammar and phonetic systems are more like those of West African languages.

Professor: Thanks for bringing that up. Most linguists agree that Gullah's origin is a practical response to the problem of communication among slaves from different tribes and countries with no common language. In fact, and this is interesting, it's speculated that the word *Gullah* may have come from the name of the Angolan tribe that was identified as "N'gola" or "Gullahs." Still in question is whether the language derived from West African Pidgin English, which the slaves may have been familiar with before they were brought to the Americas, or whether it sprang to life on the rice plantations of Georgia, South Carolina, and Florida. So that's a mystery for us to unravel! We really don't know . . .

CD 1, Track 16 PAUSE

Practice 15

First, listen to the lecture and take notes. Then identify the tone as interested, reserved, or encouraging. Compare your notes with the example notes and check your answer in the Answer Key on page 330.

CD 1, Track 16 RESUME

Notes

Tone: ______________________________

★**Bonus:** iBT Speaking
In addition to conversations, lectures, and discussions in both the iBT and the ITP Listening sections, you can use this strategy to recognize tone and emotion in conversations in Tasks 3 and 5 and lectures in Tasks 4 and 6 of the Speaking section on the iBT.

Listening Strategy 16

Listen for speakers to change their minds

Sometimes you will hear speakers say something and then change their minds immediately. This is called a *reversal*. In the question, you will be asked what the speaker meant. The original idea—the mistake that the speaker made before the reversal—will be one of the incorrect choices. Apologetic phrases often signal a reversal. The most common phrases are listed below.

Signal Words: Reversals

Sorry, Oops, I mean, Wait, Uh oh,

Example Script

Advisor's Office

Professor: So I suggest that you take English 110 this semester and then you can take 100, sorry, 120, next semester.

Student: That sounds good. Who's teaching 110?

Professor: Well, there are several sections. Sutton, Richards, Rodgers, Freedman . . .

Student: Not Freedman.

Professor: Okay. How about Richards? Did I say Richards? I meant Rodgers.

Rodgers has excellent teaching evaluations, and he is very organized.

Student: I like an organized class. You know, a class that is more . . . structured.

Professor: That should be good, then.

Student: Is Rodgers a TA though? Last semester I had a TA for my English class and frankly, she wasn't very interested in our class, or at least, it seemed like she wasn't. She was always working on the research for her dissertation and . . .

Professor: Well, most of the undergraduate writing classes are taught by TAs so . . .

Student: Oh.

Professor: And most of the professor's sections are full by the end of preregistration. But anyway, let's look. Maybe one of Townsend's sections is still open. No . . . wait, here's one. It's at four-thirty on Tuesdays and Thursdays. Would that work for you?

Student: I think so. Let me check my schedule . . . Yes, that would be great...Oops. Not Thursdays. I have a lab then.

Professor: Too bad.

Student: Maybe I could change the class, I mean the lab. Then I could take the class.

Practice 16

Listen for signal words that identify a reversal. Write down the information that you hear after the signal words. Check your answers in the Answer Key on page 331.

CD 1, Track 17

Notes

1. Signal words: ______________________________

   Reversal: ______________________________

2. Signal words: ______________________________

   Reversal: ______________________________

3. Signal word: ______________________________

   Reversal: ______________________________

4. Signal words: ______________________________

   Reversal: ______________________________

5. Signal words: ______________________________

   Reversal: ______________________________

★★**Bonus:** iBT Speaking, Writing

Speakers may use reversals in conversations and lectures in the Speaking and Writing sections of the iBT as well as in the Listening sections of both the iBT and the ITP.

Listening Strategy 17

Make connections between concepts and explanations or illustrations

When you hear professors mention something that appears to be off topic, they may have a reason for referring to it. It may be an example, a comparison that illustrates the concept, or some other related reference. Think about why the professor includes the reference. Try to make a connection.

EXAMPLE SCRIPT

Fixed Wing Aircraft

Historically airplanes have been referred to as fixed wing aircraft. They are characterized by the wings, which can be either one wing for a monoplane or . . . or two for a biplane. So besides the number of wings, we also look at the wing support, which can be braced, uh, that is, rigid . . . or . . . or something called *cantilever*, which is really just a flexible wing. Okay then. The angle of the wing is also a very important characteristic, as well as the variations along the wing and, of course, the shape, which we will talk about a little later in more detail. And, uh, most fixed-wing aircraft have a tail unit, uh, with vertical, and often horizontal, stabilizers.

Now a fixed-wing airplane is heavier than air, so how does it propel itself? And how does it stay airborne? Well, since the wings don't move, like, let's say <u>a bird . . . a bird that flaps its wings in order to stay up . . . and for the bird the flapping motion maintains . . . the flapping propels the bird forward and it also keeps the bird in the air, I mean the flapping produces lift.</u> Well an airplane doesn't rely on the motion of the wings for lift. It's actually the forward motion of the airplane that keeps it up. Here's how it works...Lift is produced when lower pressure is exerted on the upper surface of an airplane's wing compared to the pressure on the wing's lower surface. That causes the wing to be lifted up. So now let's get back to the shape of the wing. The special shape of the airplane wing . . . the airfoil . . . so that special shape is designed so air flowing over it will have to travel a greater distance and also, uh, faster, and that creates a lower pressure area that lifts the wing up. To put that another way, lift is the force that opposes the force of gravity. Another thing to remember is that the engine reduces drag, that is, resistance to the forward motion of the airplane, but the engine doesn't create lift . . .

Practice 17

Listen for the topic of a lecture and references that seem off topic. Why does the professor mention the reference or references? Compare your answers with those in the Answer Key on page 332.

CD 1, Track 18

Notes

Topic: ______________________________

References: ______________________________

Reason for the references: ______________________________

____

____

____

★★**Bonus:** iBT Speaking, Writing
The lectures in Speaking Tasks 4 and 6 as well as the lecture in the Integrated Writing Task on the iBT may include references like those in this strategy for the Listening sections of the iBT and the ITP.

Listening Strategy 18

Classify the functions of speech in replays

A few of the questions allow you to listen again to part of the conversation or lecture. When you hear the selected part the second time, you should try to identify the function of the sentences and questions that you hear. Some of the most common examples of functions are agreement, apology, assumption, complaint, disagreement, interest, refusal, regret, request, and suggestion.

Example Script

Replays

Replay: I appreciate your help, but maybe I should think about it some more.
Function: Polite refusal

Replay: So how about a group project? OR Why don't you do a group project?
Function: Suggestion

Replay: You'll never guess the outcome of the experiment. OR So what do you think happened next?
Function: Interest (to maintain attention)

Replay: Can you make copies of the handout? OR Could you make copies of the handout?
Function: Request

Replay: It's a little warm in this room. OR Do you think it's warm in here?
Function: Request (to turn on the air conditioner)

Replay: The light is on in her office.
Function: Assumption (She must be working late.)

Replay: I was surprised that my grade was so low on the exam.
Function: Complaint

Replay: I'm sorry that you didn't get a higher grade.
Function: Regret but not apology

Replay: Do you really think so? I'd be surprised at that.
Function: Disagreement

Replay: I couldn't agree with you more.
Function: Strong agreement

Practice 18

First, listen to a short exchange between two speakers. Then identify the function. Check your answers in the Answer Key on page 333.

CD 1, Track 19

1. Function: ______________________________

2. Function: ______________________________

3. Function: ______________________________

4. Function: ______________________________

5. Function: ______________________________

6. Function: ______________________________

7. Function: ______________________________

8. Function: ______________________________

9. Function: ______________________________

10. Function: ______________________________

★**Bonus:** iBT Speaking
You can use this strategy to identify functions in conversations, lectures, and discussions on both the iBT and ITP Listening sections as well as on the iBT Speaking section.

Listening Strategy 19

Recognize cues for the organization of a lecture

It is useful for you to recognize the organization of a lecture as soon as possible because it will help you to think ahead and take better notes. Professors generally plan their lectures, using the following types of organization: definition and description; chronology and process; comparison and contrast; classification; and cause and result. A topic sentence may preview the organization, but sometimes you must use cues from the lecture to identify the organization. Remember that you may hear several kinds of cues because more than one type of organization can appear in a long lecture.

Organization	*Cues*
Definition and description	*BE* verbs and descriptive adjectives
Classification	Types, parts, and kinds
Comparison and contrast	Similarity, difference, and comparative adjectives
Chronology and process	Transition words *next*, *then*, *after* or dates and steps
Cause and result	*Because* and *therefore*

Example Scripts

Definition

Fossils are the remains of ancient plants and animals that are preserved for thousands or even millions of years, usually in stone. They include animal bones, shells, and seeds, or the imprint of more fragile plant life that has disintegrated but has left its mark in a harder substance. Even footprints are fossilized.

Classification

In general, two types of planets have been identified in our solar system. The inner planets, which are mostly made of rock, are referred to as the terrestrial planets, uh, probably because of their similarity to Earth. The outer planets, those farther out in space, those have a very different kind of composition. They are, in effect, large bodies of liquid gas.

Comparison and Contrast

Unlike the direct ancestors of modern humans, Neanderthals were shorter and smaller. There is evidence that they had barrel chests and thicker bones, which may have meant that they were stronger than humans. Another significant difference was the skull and dental structure.

Chronology and Process

In 1534, Jacques Cartier sailed into the Gulf of Saint Lawrence. The next year, he traveled up the Saint Lawrence River in search of the Northwest Passage. Although he did not find it, by 1600, trade had been established with the Huron Nation. Three years later, Samuel de Champlain arrived with a view to exploring and mapping the territories in Eastern Canada, which he named New France.

Cause and Result

If lightning strikes a building with a lightning rod on it, the lightning will be conducted through the rod and into the ground through a wire instead of into the building itself. The rod attracts lightning because the air near the rod becomes ionized during an electrical storm and therefore more conductive relative to the surrounding environment.

Practice 19

Listen for the organization of the lecture. Use the topic sentence and rhetorical cues to identify the organization. Check your answers with those in the Answer Key on page 334.

CD 1, Track 20

Notes

Topic Sentence: ______________________________

Cues: ______________________________

Organization: ______________________________

★★**Bonus:** iBT Reading, ITP Reading
You can also use this strategy to identify the organization of reading passages on both the iBT and the ITP.

Listening Strategy 20

Use lists and charts for notes

You will have to arrange events in order or fill in a chart that shows relationships in some passages. Comparison and contrast are easy to arrange on a chart and are the most common organization for a lecture that will include a relationship question. If you use lists and charts in your notes, you will have a visual reference for the relationships in the lecture. It will be easy for you to answer these questions.

Example Script

Humanistic Therapies

All three of the humanistic therapies that we are going to discuss today are generally . . . let's say, optimistic. They tend to . . . to support the clients and help them to fulfill their potential. Carl Rogers developed what he called Client-Centered or Person-Centered therapy that, uh, encourages the therapist to focus on the client's point of view instead of the therapist's interpretation. So, logically then, the therapist is very nonjudgmental and doesn't really provide subjective direction for the clients. Active listening on the part of the therapist allows the clients to share both the information in the communication and the emotion that they feel. One of the most useful techniques in Client-Centered therapy is *mirroring*, uh, that means that the therapist simply repeats a summary of what the clients say in sessions so they can hear their own words and reflect on them.

Okay, the next therapy, Gestalt therapy, that's also a humanistic approach, but its goal is to help clients to become more integrated, as the name would imply. Fritz Perls, who established Gestalt therapy, he thought that it was important to help clients accept and integrate painful or undesirable aspects of their personalities. To do that, he preferred to hold sessions in groups. Although the emphasis was still on the individual . . . the potential of the individual . . . he thought that working out problems with the support of a group would help individuals to assume more personal responsibility. All communication between the therapist and the group in a Gestalt session is in the present continuous tense because what is happening *now* is primary. The interpretation of dreams, fantasy, and role playing are also important techniques for a Gestalt therapist.

So that brings us to Existential therapy, that's a humanistic approach too, but it begins with larger questions about the meaning of life and assists clients to . . . to . . . to develop a value system. The goal of this therapy, according to Victor Frankl, uh, the goal is to help clients understand the freedom of choice that is available to them and to find out what is important to them so they can find purpose and meaning in life . . .

EXAMPLE NOTES

Client-Centered	**Gestalt**	**Existential**
Carl Rogers	Fritz Perls	Victor Frankl
client point of view	accept painful/undesirable	larger questions
nonjudgmental	group sessions	meaning of life
active listening	role playing/dreams	value system
mirroring—repeat summary	NOW—present continuous tense	choices—purpose

EXAMPLE QUESTION

With which type of therapy are the following techniques associated? Please place a check mark in the correct box.

	Client-Centered	Gestalt	Existential
group sessions		✔	
mirroring	✔		
value system			✔
present continuous tense		✔	
active listening	✔		

Practice 20

Listen for sequences and relationships in the lecture. Write notes in the form of lists and charts. Compare your lists or charts with those in the example notes and the example question in the Answer Key on page 336.

CD 1, Track 21

Notes

★★**Bonus:** iBT Speaking, Writing
You can also use this strategy when you are taking notes for conversations and lectures in the Speaking and Writing sections of the iBT. Remember that you cannot use notes on the ITP.

4

STRATEGIES FOR THE SPEAKING SECTION

OVERVIEW OF THE iBT SPEAKING SECTION

The Speaking section is Section 3 on the Internet-Based TOEFL. The Speaking section tests your ability to communicate in English in an academic setting. During the test, you will be presented with six speaking questions. The questions ask for a response to a single question, a conversation, a talk, or a lecture.

You may take notes as you listen, but notes will not be graded. You may use your notes to answer the questions. Some of the questions ask for a response to a reading passage and a talk or a lecture. The reading passages and the questions are written, but the directions will be spoken.

Your speaking will be evaluated on both the fluency of the language and the accuracy of the content. You will have 15–20 seconds to prepare and 45–60 seconds to respond to each question. Typically, a good response will require all of the response time, and the answer will be complete by the end of the response time.

You will have about 20 minutes to complete the Speaking section. A clock on the screen will show you how much time you have to prepare each of your answers and how much time you have to record each response.

A Speaking section is NOT included in the ITP TOEFL.

> When you practice the Speaking strategies, you will be asked to record some of your answers so that you can compare them with the example answers in the Answer Key in Chapter 7. Do this on your phone, iPod, or another recording device. You cannot record on the CDs included in this book.

TIPS FROM SUCCESSFUL TOEFL STUDENTS

WHAT TO DO ON THE SPEAKING SECTION

- Don't worry about your accent. Everyone who takes the TOEFL has an accent! Just try to speak so that the raters can understand you.
- Try to make friends with English-speaking people. Join clubs or participate in activities where you will meet people who speak English. Ask them their opinions and listen carefully to the way that they talk to you.
- Record yourself while you are practicing. After you listen to your first answer, practice the same question again and record yourself. Compare the two recordings.
- Adjust the microphone before you start to speak. Then be careful not to touch the microphone while you are speaking because it can cause noise that will bother the raters.
- It's okay to make a few mistakes when you are speaking. You can still get an excellent score on the Speaking section without submitting six perfect responses.

Speaking Strategy 1

Use the break to prepare for the Speaking section

Use the 10-minute break required at the end of the Listening section to get ready for the Speaking section by talking in English. Do not speak your language with friends who may also be taking a break. Speak English with the test administrator. Ask if it is okay to speak English with other people. Tell the administrator that you would like to warm up for the next section of the test by speaking English. If the administrator says that you cannot speak with someone else, then talk to yourself in English, either quietly aloud or silently in your mind.

EXAMPLE CONVERSATIONS

Student: Excuse me. May I speak English with other students during the break? I want to get ready for the Speaking section of the test by speaking English.

Test Administrator: Sure. Just so you don't disturb anyone else. And be sure to be back here to sign in before the 10-minute break is over.

Student: Okay. Thank you.

Test Administrator: No problem.

Student: Excuse me. May I speak English with other students during the break? I want to get ready for the Speaking section of the test by speaking English.

Test Administrator: Sorry. No talking allowed.

Student: Okay. Thanks anyway.

Practice 1

CD 2, Track 1

Prepare to ask the administrator whether you are allowed to speak English during the break. First plan and write a question or an answer to each of the five situations below. Then compare your answers with the examples in the Answer Key on page 337.

1. Ask the administrator's permission to speak English with others during the break.

2. Explain that you want to speak English because you are preparing for the Speaking section after the break.

3. Thank the administrator for allowing you to speak English during the break.

4. Thank the administrator for NOT allowing you to speak English during the break.

5. Ask the administrator for directions to the bathroom or a water fountain.

Even if you cannot speak English with others, you can continue thinking in English. Have a positive conversation in English in your mind. This is a good way to prepare for speaking aloud!

Speaking Strategy 2

Stay positive and confident

Before you record the questions on the six tasks, you will speak into the microphone to adjust the volume. You will be asked a very easy question, for example: *Describe a town that you know well*. This is your opportunity not only to adjust the volume of the computer but also to adjust your attitude. It is important to sound confident, even if you don't feel confident. To do this, you need to have a plan.

Example Plan

1. Take deep breaths! Breathe in and out at least three times. As you breathe in, affirm your success in your mind. Tell yourself "I am ready" as you breathe in. Tell yourself "I feel confident" as you breathe out.
2. Smile! Smile at the computer screen. It will make you feel better, and it will exercise the muscles around your mouth. Continue to smile as you begin speaking.
3. Speak up! A shy whisper doesn't sound very confident, and even though the raters are not supposed to be influenced by the quality of your voice, we all know that the raters have to hear you in order to evaluate you. A confident voice can make the difference when the raters are deciding between a lower and a higher number.
4. Keep going! If you think that you have not scored well on Task 1, don't look back. Try to do your best on Task 2. Concentrate on the task in front of you.

Stay positive! Some people take the TOEFL more than once, but they reach their goals. Make friends with positive people and encourage each other.

Example Answer

The place where I live now is Flagstaff, which is a college town in Northern Arizona. It's a very beautiful place because it's surrounded by high mountain peaks that usually have snow on them, even in the summer. The hills nearby are covered with pine trees that are dark green year round. The town itself has the usual businesses that surround a campus and cater to students, but in Flagstaff, many of them are built in the style of an old western settlement. Some of the stores sell jewelry, pottery, and rugs that are supplied by the Navajo and Hopi people who live nearby. I like Flagstaff because there's a lot to do. Besides shopping, there are a lot of cultural activities on campus. If you like being outdoors, you can go skiing in the Snow Bowl or hiking on one of many trails. You can even go camping in the Grand Canyon, only two hours away.

Practice 2

CD 2, Track 2

Practice using the plan when you answer the volume question. Speak for about a minute. Breathe, smile, speak up, and keep going! Be sure to record your response and compare it with the example answer in the Answer Key on page 337.

Volume Question: Describe a town or city that you know well. Be sure to include examples and details.

★★★★**Bonus:** Listening, Structure, Reading, Writing
It is important to maintain a positive attitude as you progress through all sections of the TOEFL.

Speaking Strategy 3

Get used to hearing others

You should expect to hear others speaking while you are listening to the questions, preparing your responses, and recording the speaking tasks on the iBT TOEFL. To simulate the test situation, turn on a radio, TV, or other audio device while you are taking model tests to practice the Speaking section. Be sure to select a program that features speech, not music. Learn to ignore the noise.

EXAMPLE

Integrated Speaking Task 4

First read the passage on hibernation and then listen to a lecture on the same topic. Finally, listen to a typical question and response to the speaking task in a setting where several people are taking the iBT TOEFL.

READING TIME: 45 SECONDS

Hibernation

Hibernation is an inactive state in which the metabolic system in animals is depressed. The temperature of the body decreases and the breathing slows down. Hibernation conserves energy during winter or an extremely hot summer when food is less abundant. Depending on the species, hibernation can last for days or months. Some animals experience a state of total hibernation, but typically, an inactive state is accompanied by short periods of arousal when the body temperature readjusts to normal levels. This intermittent hibernation is technically referred to as torpor; however, the term hibernation is generally used to identify both long-term and short-term periods of inactivity in animals.

EXAMPLE SCRIPT

CD 2, Track 3 PAUSE (after lecture)

Notes

R [READING]

H—metabolism reduced/temp/breathing

- Total
- Intermittent—torpor

L [LECTURE]

Frogs—cold blooded

- Bury—mud/ice
- Less oxygen—absorb skin
- Eat 1%

Practice 3

Integrated Speaking Task 4

In Question 4 you will read a short passage and listen to part of a lecture on the same topic. Then you will listen for a question about them. After you hear the question, you have 30 seconds to prepare and 60 seconds to record your answer. Be sure to record your response and compare it with the example answer in the Answer Key on page 338.

Now read the passage about phobias.

READING TIME: 45 SECONDS

> **Phobias**
> A phobia is an irrational fear that interferes with the daily routine of the person experiencing it. In a mild form, it inhibits function, but in severe and chronic conditions, it can be disabling. To qualify as a phobia, the fear must be disproportionate to the situation or experience. In effect, relatively harmless stimuli can trigger an exaggerated response in the brain. Another condition that distinguishes simple fear from a phobia is avoidance. The person experiencing a phobia will respond by avoiding the triggering situation, even when it is inconvenient or restricts normal activities. Phobias are the most common type of anxiety disorder.

Now listen to part of a lecture in a psychology class. The professor is talking about phobias. Using information from both the reading passage and the lecture, discuss phobias and give examples of them. For this practice activity, plan your response and begin speaking after the beep. Try not to become distracted by the responses from other speakers.

Preparation Time: 30 seconds Recording Time: 60 seconds

CD 2, Track 3 RESUME

★**Bonus:** iBT Listening
You can also use this strategy to learn how to focus on the passages in the Listening section of the iBT because others may begin their Speaking section before you do.

Speaking Strategy 4

Use shortcuts for notes

Your notes for the Speaking section of the TOEFL are really memory aids to help you recall what you want to say. You don't have much time to write because your preparation time is short. That is why you need to use shortcuts.

1. ***Abbreviate key words***
2. ***Use symbols***

Examples

1. ***Abbreviate key words***

Key words are repeated. When you hear a key word for the first time, you can write it out, but when you hear it again and recognize it as a key word, just capitalize the first letter.

Domesticated Animals	DA
Wild Animals	WA
Horse	H

2. ***Use symbols***

A dozen small words are in most high-frequency word lists. That means you will probably hear them often in the Speaking section. The following symbols will help you take notes more quickly. When you hear one of the high-frequency words, note it with a symbol.

+	and
w	with
w/o	without
o	or
=	is, are, means, refers to, is called
≠	not, not the same
@	about, approximately
X	example
>	more, larger
<	less, smaller
→	results in, causes, produces, therefore
←	comes from, derives from

Practice 4

First listen to the sentences and take notes. Use the shortcuts you have learned. Abbreviate key words that are repeated throughout the lectures and receive emphasis in speaking. Key words have been underlined here for practice. Use symbols. Finally, compare your answers with the example notes in the Answer Key on page 339.

CD 2, Track 4

1. Volcanoes and geysers have some common elements.

2. Mercury is not a solid.

3. Babies demonstrate understanding of mathematical concepts at about six months old.

4. Few ancient cities are constructed without walls.

5. The paintings of flowers by O'Keeffe were larger than the works of her contemporaries.

6. The oxygen supply to the cornea of the eye comes directly from the air.

7. An example of a successful online business is eBay.

8. An ornithologist is a scientist who specializes in the study of birds.

9. The architecture of many early public buildings includes a dome or a tower.

10. Raindrops refract light like a prism, causing the colors in a rainbow.

★★**Bonus:** iBT Listening, Writing
Shortcuts will also be useful when you are taking notes on the listening passages in both the Listening section and the Writing section of the iBT.

Speaking Strategy 5

Draw maps to organize notes

You will take notes on two lectures on the Speaking section. Listen for cues about the way that each lecture is organized. Draw maps to show the relationships in your notes. Some of the most common organizational patterns are listed below.

Definition
Classification
Sequence
Comparison and contrast
Cause and effect

EXAMPLES

Definition

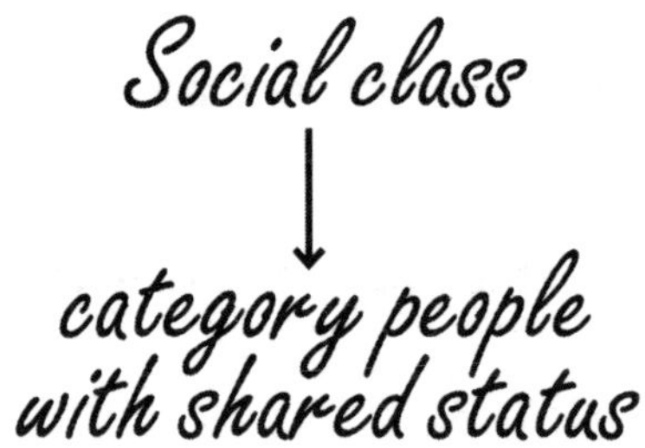

Classification

Sequence

19th C painting
- Romanticism 1800s
- Realism 1850
- Impressionism 1870

Comparison and Contrast

Conditioning

Classical	Operant
• involuntary	• voluntary
• key events evoke behavior	• key events produced by behavior
• absence key events no behavior	• absence of stimuli behavior occurs

Cause and Effect

injured cells ⟶ chemicals ⟶ pain

Practice 5

First read the sentence. Then listen to it and take notes. Use the maps you have learned. Finally, compare your answers with the maps in the Answer Key on page 340.

CD 2, Track 5

1. Migration can cause variation in populations due to the introduction of new genes.

2. The right hemisphere of the brain controls the left side of the body and contributes to spatial, artistic, and musical tasks, whereas the left hemisphere controls the right side of the body and is essential to verbal and logical tasks.

3. Somatotyping is a system that describes personality in terms of an individual's physique.

4. Historians believe that the first coins may have been created about 700 B.C.E. in Anatolia; the Greeks and Persians were exchanging coins between 650 and 500 B.C.E.; by 300 B.C.E., the Romans had already begun to mint coins and distribute them at temples.

5. The two major types of seed plants are gymnosperms and angiosperms.

★★★**Bonus:** iBT Listening, Writing, ITP Essay
Maps will be useful when you are taking notes on the listening passages in both the Listening section and the Writing section of the iBT. You can also draw maps when you are brainstorming your essays.

Speaking Strategy 6

Use a direct approach to begin your talk

On some tasks, you have 45 seconds to answer. On other tasks, you have 60 seconds to answer. You don't have time for a long introduction. You need to get to the point in the first sentence. You need to start with a direct statement.

Direct statements in English are usually Noun–Verb–Noun. They are generally short, but remember that a noun can be a clause, as, for example, "that class attendance should be optional."

EXAMPLES

Task 1: My favorite pastime is reading.

Task 2: I think that class attendance should be optional.

Task 3: The student does not favor the university's plan to build a radio station.

Task 4: The experiment proves that babies can communicate before they speak.

Task 5: The problem is that the woman's passport has expired.

Task 6: Two types of farming methods were discussed in the lecture.

Practice 6

Choose the direct statement for each task. Then read your sentences aloud. Be sure to record your response and compare your answer with the example answer in the Answer Key on page 341.

CD 2, Track 6

Task 1
There are many attractive cities in the United States that I would like to visit if I could go there.

I would like to visit Los Angeles.

Task 2
I think that traditional schools offer many advantages.

Some people think that home schools with the support of technology offer the best education for children and other people think that traditional schools are a better choice.

Task 3
The university has changed its policy regarding the number of transfer hours a student can apply to a degree program.

The student supports the new transfer program that the university has instituted.

Task 4
The lecturer's examples refute the research study on facial expressions.

I listened to a lecture and I read a paragraph about facial expressions.

Task 5
The woman has a serious problem that she explains to the man.

The woman's problem is that she has lost her student ID card.

Task 6
Three main types of volcanoes are identified in the lecture, including shield volcanoes, cinder cone volcanoes, and composite volcanoes.

Volcanoes are a very interesting topic that the lecturer presents in this question.

★★**Bonus:** iBT Writing, ITP Essay
This strategy is also a good way to begin an essay; however, on an essay, you have more time and you can include a longer introduction.

Speaking Strategy 7

Give your opinion if you are asked for it

Listen carefully to the directions. Some of the speaking tasks ask for your opinion, but some of the tasks ask you to report someone else's opinion, or simply to summarize facts. Give your opinion if you are asked for your opinion. Report the opinions of the speakers if you are asked to report their opinions. Summarize the facts if you are not asked for an opinion.

EXAMPLES

Task 1
Choose a person whom you admire and explain why this person deserves your admiration. Use specific details and examples to support your response.
Your opinion.

Task 2
Some people like to vacation at the beach. Other people prefer to go to a city on their vacations. Where would you rather go and why? Use specific reasons and details to support your response.
Your opinion.

Task 3
The student expresses her opinion about the change in the final examination schedule. State her opinion and explain her reasons for having that opinion.
The student's opinion.

Task 4
Explain how the example in the professor's lecture casts doubt on the concept of kinship in the reading passage.
Facts without opinion.

Task 5
Explain the man's problem and the two possible solutions that the woman suggests. Then state your preference and explain why.
Your opinion about the best solution.

Task 6
Using the ideas and examples from the lecture, explain how the international space station has contributed to both science and intercultural relationships.
Facts without opinion.

Practice 7

First read the question. Then determine which of the following the question requires: your opinion, the report of another person's opinion, or facts without an opinion. Compare your answers with those in the Answer Key on page 341.

1. Some people think that employees should be paid based on achievements and merit. Other people think that seniority, that is, the length of time that they have been employed, is more important. What do you think and why?

2. Explain the student's problem and the possible solutions that the professor recommends. What do you think that the student should do and why?

3. Using the ideas and examples from the lecture, explain the three types of nonverbal behavior that the professor mentions in the lecture.

4. The man expresses his opinion about the professor's policy for class participation. State his opinion and the reasons he has for having that opinion.

5. Describe an ideal job. Use specific reasons and details to support your response.

6. Explain how the research study that the professor cites in his lecture supports the use of clean coal as an alternative energy source.

7. Explain the woman's problem and the two possible solutions that her friend suggests. Then choose the better option and justify your choice.

8. Some people believe that it is better to marry young. Other people advise young people to wait until later in life to marry. Which idea do you think is better?

9. Using the main ideas and supporting examples from the lecture, explain how laser beams work and how they support modern products and technologies.

10. The man expresses his opinion about the requirement that his advisor sign all of his course requests before registration. State his opinion and the reasons that he gives for holding that view.

★★**Bonus:** iBT Writing, ITP Essay
Determining whether to give personal opinions is a good strategy for the Writing section of the iBT as well as for the Speaking section. The Writing section of the ITP is an opinion essay.

Speaking Strategy 8

Be sure of your vocabulary and grammar choices

Some words and grammar are more challenging for you. If you always have pronunciation problems with certain sounds, choose alternative words and phrases that do not have that sound in them. If you always make grammatical mistakes using certain structures, choose alternative structures that you are sure of. Although it is good to try to improve your vocabulary and grammar before the test, when you have to speak, use the alternatives that you have mastered before the test. Raters will not be impressed by big words that you cannot pronounce correctly or complex grammar that is not used correctly. Of course, sometimes it is necessary to use a challenging word or structure on the test. In that case, do your best and go on.

EXAMPLES

l/r sounds

Challenging	***Possible Alternative***
I believe	In my opinion
Personally	It seems to me
Generally	For the most part
In conclusion	To sum up
Clearly	Without a doubt
Recommend	Suggest
Rather	Prefer to
Lecturer	Speaker

v/w/b sounds

Challenging	***Possible Alternative***
Very	Quite
Valid	Correct
Variables	Factors
Verify	Check
Virtually	Almost all
View	Opinion

th/s sounds

Challenging	***Possible Alternative***
I think	I believe
Third	Last
Thought	Idea
For one thing	In the first place
Theory	Assumption
Method	Procedure

Practice 8

First listen to the sentence and try to repeat it. Be sure to record your response and compare it with the example answer in the Answer Key on page 342. Then decide whether you need an alternative. If you are satisfied with your pronunciation, mark the sentence with a check. If you feel that one of the words is challenging, use an alternative.

CD 2, Track 7

1. The lecturer describes two types of competition.
2. I believe that the man should join a study group.
3. A number of variables should be taken into consideration.
4. For one thing, the parking lots are always full.
5. In my view, students should live on campus.
6. I'd rather go to school at night than during the day.
7. Virtually every student signed the petition.
8. Personally, I think that the steam engine was the most important invention.
9. The professor makes a very good case for changing the schedule.
10. I think that grades should be based on tests, not on class discussions.
11. The reading passage summarizes the theory of plate tectonics.
12. It's a valid argument.
13. Third, exercise is essential.
14. The example in the talk clearly explains the concept.
15. The man's friend recommends that he take a break.
16. The research was verified by a second study.
17. The lecturer disagreed with the method.
18. Generally speaking, mathematics is the study of time, quantity, and distance.
19. In conclusion, insects adapt to survive.
20. Several thoughts were presented.

★★**Bonus:** iBT Writing, ITP Essay
Although it is not necessary to pronounce words when you write an essay, it is still a good plan to choose vocabulary and grammar that you know how to use.

Speaking Strategy 9

Acknowledge the other opinion with a concession

Concession means to show respect for a different opinion without agreeing. After the concession, you express your opinion. Acknowledging an alternative opinion is a traditional option for presenting your opinion. The language of concession includes the following words and phrases: *although*, *even though*, *despite*, *in spite of*, *but, however*, *nevertheless*.

Task 2

Some people like living in a city. Other people prefer living in a small town. Where do you prefer to live, and why? Be sure to use details and examples to support your opinion.

EXAMPLES

Although living in a city has many advantages, I prefer to live in a small town.

Even though living in a city has many advantages, I prefer to live in a small town.

Despite many advantages of living in a city, I prefer living in a small town.

In spite of many advantages of living in a city, I prefer living in a small town.

Living in a city has many advantages, but I prefer living in a small town.

Living in a city has many advantages; however, I prefer living in a small town.

Living in a city has many advantages; nevertheless, I prefer living in a small town.

Although living in a city has many advantages, I would rather live in a small town.

In spite of many advantages of living in a city, I would rather live in a small town.

Living in a city has many advantages, but I would rather live in a small town.

Practice 9

First read the task. Next write concession sentences to respond to the task. Fill in the blanks to complete the sentences. Then read your sentences aloud. Be sure to record your responses and compare them with the example answers in the Answer Key on page 343. Finally, without looking at the sentences you wrote, practice saying some of the concession sentences from memory.

CD 2, Track 8

Task 2

1. Some students want a room in a dormitory. Other students want to live in an apartment. Which type of living situation do you prefer, and why? Use specific reasons and examples to support your opinion.

   Although ____,

   I prefer ____.

2. Some people like to take tours. Other people prefer to travel alone. Which experience would you prefer, and why? Use specific reasons and examples to support your opinion.

   Even though ____,

   I prefer ____.

3. Some people want a job that provides them with a secure income in a company. Other people want to take a risk and go into business for themselves. Which type of job do you prefer, and why? Use specific reasons and examples to support your opinion.

   Despite ____,

   I prefer ____.

4. Some people like to use e-mail and texting even for important communications. Other people prefer to conduct important communications by telephone or in person. Which methods do you prefer, and why? Use specific reasons and examples to support your opinion.

   In spite of ____,

   I prefer ____.

5. Some students want to live alone. Other students want to live with a roommate. Which type of arrangement do you prefer, and why? Use specific reasons and examples to support your opinion.

   ____,

   but I prefer ____.

6. Some people believe that you should never tell a lie. Other people think that it is better not to tell the truth when it will hurt someone else's feelings. What do you think, and why? Use specific reasons and examples to support your opinion.

   ____; however,

   I prefer ____.

7. Some people enjoy cultural activities. Other people would rather go to sports events. Which kind of activity do you prefer, and why? Use specific reasons and examples to support your opinion.

   ____; nevertheless,

   I prefer ____.

8. Some people like to see serious dramas. Other people would like a comedy when they choose a movie. Which type of movies do you prefer, and why? Use specific reasons and examples to support your opinion.

   Although ____,

   I would rather ____.

9. Some students think that campus activities take too much time. Other students like to join student clubs to make friends. Which decision would you make, and why? Use specific reasons and examples to support your opinion.

   In spite of ____,

   I would rather ____.

10. Some people like to have a lot of friends. Other people want only a few good friends. Which choice do you support, and why? Use specific reasons and examples to support your opinion.

    ____, but

    I would rather ____.

★★**Bonus:** iBT Writing, ITP Essay

Concession is a useful strategy for the opinion essay in the Writing sections of both the iBT and the ITP as well as for Task 2 in the iBT Speaking section.

Speaking Strategy 10

Support your answers with reasons

You must support your answer on every task that asks for an opinion or preference. Practice giving an opinion and the reasons why you have that opinion. Start using the word *because* after every opinion and then be sure to include two or three reasons for the opinion. If you include more than two or three reasons, you won't have time to explain them or provide details in the limited time you have to answer the question.

Grammar Reminder

Use a subject and verb after *because*:

I like traveling because I meet interesting people.

Use a noun or noun phrase after *because of*:

I like traveling because of the interesting people.

EXAMPLES

I don't think that college students should have jobs because working takes time away from studying and it prevents students from participating in campus activities.

My favorite movies are comedies because laughing is a good way to relax and comedies usually aren't as violent as action movies.

My hometown is an interesting place because of the history and the scenic beauty.

The privacy of famous people should be respected because they need some time to interact alone with their families and they deserve the same rights to privacy that ordinary people enjoy.

In my opinion, computer courses are better than traditional courses because of the convenience and the opportunity for individualized instruction.

Practice 10

First read the question. Next write an opinion sentence to respond to the task. Be sure to include reasons for your opinion. Check your grammar. Then read your sentences aloud. Be sure to record your responses and compare them with the example answers in the Answer Key on page 344. Finally, without looking at the sentences you wrote, give your opinion for each question.

Tasks 1 and 2

1. What is your favorite place to go when you have free time?

2. Do you think that it is better for a family to live in one place or move to different places?

3. In your opinion, what is the most important skill or lesson you learn in college?

4. Some people believe that they can succeed by following a plan while other people believe that they will succeed by taking advantage of chance opportunities. What is your opinion?

5. Who is your most important advisor?

6. The university can use a grant to build a sports center or a branch library. Which project do you favor?

7. Would you rather own a business or work for a company?

8. Should men and women attend the same school or separate schools?

9. If you received a gift of one thousand dollars, what would you do?

10. Some people like to make new friends and other people prefer to spend time with friends that they have known for a long time. What do you prefer to do?

★★**Bonus:** iBT Writing, ITP Essay
Supporting opinions is an important strategy for the opinion essay in the Writing sections of both the iBT and the ITP as well as for Task 1 on the iBT Speaking section.

Speaking Strategy 11

Credit the ideas of others with a citation

Citation means to give someone else credit for an idea. You will lose points on the Speaking section for repeating sentences without citing them. Three ways to use citation are to introduce the source, to use strong verbs to report ideas, and to use the 1-2-3 reference.

Introduce the source	***Use strong verbs***	***Cite with the 1-2-3 reference***
According to . . .	argues	First citation: full name
the reading	concludes	Second citation: last name
the study	maintains	Third citation: pronoun
the professor	mentions	
the lecturer	points out	
Dr. Adams	proposes	
	states	
	suggests	

EXAMPLES

According to the reading, biologists have traditionally divided bacteria into two types—helpful and non-helpful bacteria.

According to the lecturer, the classifications may need to be revised in light of new evidence.

According to the reading passage, probiotics are helpful bacteria that colonize the gastrointestinal tract.

The professor points out that all probiotics may not be helpful.

Dr. Thomas states that bacteria may be either helpful or non-helpful. [First reference]

Thomas mentions that probiotics are helpful bacteria that are usually found in the gastrointestinal tract. [Second reference]

He concludes that not all probiotics are helpful, however. [Final reference]

Practice 11

First read the information. Then, using the source listed, write a citation and read your answers aloud. In 1–5 you will need one sentence. In 6–10, you will need three sentences. Be sure to record your responses and compare them with the example answers in the Answer Key on page 345.

1. Reading passage
   The jaw of the great white shark is lined with rows of replacement teeth.

2. The lecturer
   Only 2 percent of Antarctica is free of ice.

3. The professor
   Secondary colors are made by combining two primary colors.

4. Reading
   Estuaries are habitats for diverse wildlife.

5. The study
   The blue whale can dive approximately 100 feet, which is about 250 meters.

6. Anthropologist George Wharton James
   The Hopi people have traditionally made pottery and woven baskets.
   Bright colors such as red, green, and yellow are typical of the designs.
   Either geometric forms or ceremonial Kachina figures dominate the patterns.

7. Professor Abdul Latiff Mohammad
   The *Rafflesia* is a parasitic flowering plant found in the rainforests of Indonesia, Malaysia, and the Philippines.
   It takes some species as long as six to nine months to complete the budding period.
   A single bloom of the *Rafflesia* plant can grow to more than 3 feet long and weigh over 15 pounds.

8. Dr. William Fry
   Laughing seems to provide the same benefits as exercise.
   Laughter increases the release of endorphins in the body.
   When patients laugh, they require less pain medication.

9. Dr. Carl Sagan
   Life on Earth is not exceptional.
   Life probably exists elsewhere in such a vast universe.
   Consistent physical laws increase the probability of finding extraterrestrial life.

10. Dr. Edward Hall
    65 percent of the meaning in a normal conversation is transmitted through nonverbal cues.
    People from different cultures use gestures in different ways.
    Sometimes a harmless gesture in one culture is interpreted as offensive in another culture.

★**Bonus:** iBT Writing
Citations for information from other sources are necessary for the integrated essay in the Writing section of the iBT.

Speaking Strategy 12

Use the directions as an outline

Before you begin each task, you will hear directions. Listen carefully! The directions are an outline of what you should include in your answer. If you follow the directions, you will include all the parts of your talk and stay within the time limits without being cut off by the beep. When you are asked for specific reasons and details, remember that you only have time for two or three. The answer is really in the question.

EXAMPLES

Read the directions and compare them with the preparation notes. Notice how the question serves as an outline for each task.

Task 1

Talk about your favorite subject in school. Include specific reasons and details to explain your choice.

Preparation Notes

Favorite subject

Reasons

-
-
-

Task 2

Some people like to be engaged in several tasks at the same time. Other people prefer to focus on one task at a time. Which do you prefer, and why? Use specific reasons and examples to support your opinion.

Preparation Notes

Concession

Choice

Reasons

-
-

Conclusion

Task 3
Read an announcement from the campus newspaper about a new program at the university.

READING TIME: 45 SECONDS

Now listen to two students who are talking about the new Child Development Center at the university. The woman expresses her opinion of the program. Report her opinion and explain the reasons that she has for having that opinion.

Preparation Notes

Announcement

Opinion

Reasons

-
-
-

Task 4
Read the passage about nonverbal communication.

READING TIME: 45 SECONDS

Now listen to part of a lecture in a linguistics class. The professor is talking about nonverbal communication. Using information from both the reading passage and the lecture, discuss nonverbal communication and provide examples from the research study.

Preparation Notes

Reading Concept

Transition

Lecture Examples

-
-
-

Conclusion

Task 5

Describe the man's problem and the solutions that his friend suggests. What do you think the man should do, and why?

Preparation Notes

Problem

Suggestions

-
-

Opinion

Reasons

-
-

Task 6

Using the main points and examples from the lecture, discuss research in the canopy of the rainforest.

Preparation Notes

Main Idea 1

Main Idea 2

-
-

Main Idea 3

-
-

Practice 12

First listen to the response for each task and fill in the outline for the preparation notes. Then compare your outlines with the speaker's preparation notes in the Answer Key on page 346.

CD 2, Track 9

Task 1
Talk about your favorite subject in school. Include specific reasons and details to explain your choice.

<u>**Preparation Notes**</u>

Favorite subject

Reasons

•
•
•

Task 2
Some people like to be engaged in several tasks at the same time. Other people prefer to focus on one task at a time. Which do you prefer, and why? Use specific reasons and examples to support your opinion.

<u>**Preparation Notes**</u>

Concession

Choice

Reasons

•
•

Conclusion

Task 3
Read an announcement from the campus newspaper about a new program at the university.

READING TIME: 45 SECONDS

CD 2, Track 9 PAUSE

> **Child Development Center**
> East Campus will be opening the doors of the new Wilson Child Development Center in September. The purpose of the 12-million-dollar facility is twofold. In partnership with the College of Education, the Child Development Center will serve as a laboratory workshop for students in the Early Childhood program at the University. In addition, 150 children of faculty and students will be enrolled in an advanced curriculum for three to five-year-olds. Both half-day and full-day options are still available. To find out about the laboratory workshop, please contact the College of Education on the main campus. To enroll you child, please visit the Child Development Center on East Campus.

CD 2, Track 9 RESUME

Now listen to two students who are talking about the new Child Development Center at the university. The woman expresses her opinion of the program. Report her opinion and explain the reasons that she has for having that opinion.

Preparation Notes

Announcement

Opinion

Reasons
-
-
-

Task 4
Read the passage about nonverbal communication.

READING TIME: 45 SECONDS

CD 2, Track 9 PAUSE

Nonverbal Behavior
People respond strongly to the nonverbal behavior that accompanies speech. Nonverbal behavior can include not only facial expressions, eye contact, and gestures but also the amount of space between the speaker and listener and the tone of voice. In some cases, touch can also carry a message. Traditionally, it has been assumed that between 70–80 percent of a communication is transmitted by nonverbal behavior. In fact, a more recent study at UCLA found percentages as high as 93 percent. Clearly, when there is a mismatch between the verbal and the nonverbal message, what we say will be less likely to be believed than the way in which the nonverbal message was communicated.

CD 2, Track 9 RESUME

Now listen to part of a lecture in a linguistics class. The professor is talking about nonverbal communication. Using information from both the reading passage and the lecture, discuss nonverbal communication and provide examples from the research study.

Preparation Notes

Reading Concept

Transition

Lecture Examples
-
-
-

Conclusion

Task 5

Now listen to a short conversation between a student and his friend. Describe the man's problem and the solutions that his friend suggests. What do you think the man should do, and why?

Preparation Notes

Problem

Suggestions

-
-

Opinion

Reasons

-
-

Task 6

Now listen to part of a lecture in an environmental science class. Using the main points and examples from the lecture, discuss research in the canopy of the rainforest.

Preparation Notes

Main Idea 1

Main Idea 2

-
-

Main Idea 3

-
-

Speaking Strategy 13

Memorize a few key phrases

It is NOT a good idea to memorize answers from the strategies in this book or from any other resources. But it IS a good idea to have some key phrases in mind for each of the tasks. That way, if you become very nervous, you will be able to get back on track.

EXAMPLES

Task 1

_________________ is my favorite _________________.

_________________ is the place that _________________.
person
object
event
activity

In my experience, _________________.

In my view, _________________.

If _________________ could _________________, _________________ would _________________.

Task 2

Although _________________ is _________________, I prefer _________________.

Although a case could be made for _________________, I _________________.

In my experience, _________________ because _________________.

I think that _________________ is better than _________________ because _________________.

I agree that _________________ because _________________.

I prefer _________________ to _________________ because _________________.

Task 3

According to ________________, ________________.

The man objects to ________________ because ________________.

He favors ________________.

He presents three arguments against ________________.

The woman agrees that ________________.

The woman agrees with the policy. She explains that ________________.

The woman thinks that ________________ is a good idea. She points out that ________________.

Task 4

A(n) ________________ is ________________ that ________________.

According to the ________________, ________________.

The lecturer ________________ the information in the reading.

Task 5

The problem is that ________________.

According to ________________, one solution is to ________________.

Another possibility is to ________________.

I think that the best solution is to ________________ because ________________.

In my opinion, the man/woman should ________________.

Task 6

According to the lecturer, _________________.

The lecturer states that _________________.

The two/major _________________ are _________________.

The lecturer concludes that _________________.

In other words, _________________.

Practice 13

First listen to the speakers as they fill in the blanks, using the key phrases. Then think about how you would fill in the blanks for each sentence and fill in the blanks using your own information. Be sure to record your responses and compare them with the example answers in the Answer Key on page 352.

CD 2, Track 10

Task 1

______________________ is my favorite ______________________.

______________________ is the place that ______________________.
person
object
event
activity

In my experience, ______________________.

In my view, ______________________.

If ____________ could ____________, ____________ would ____________.

Task 2

Although ______________________ is ______________________,

I prefer ______________________.

Although a case could be made for ______________________, I ______________.

In my experience, ______________________ because ______________________.

I think that ____________ is better than ____________ because ____________.

I agree that ______________________ because ______________________.

I prefer ____________ to ____________ because ____________.

Task 3

According to ______________________________, ______________________________.

The man objects to __________________________ because ________________________.

He favors ____.

He presents three arguments against ____.

The woman agrees that ____.

The woman agrees with the policy. She explains that __________________________________.

The woman thinks that ____ is a good idea.

She points out that ____.

Task 4

A(n) ____________________ is _____________________ that _____________________.

According to the ____________________________, ___________________________.

The lecturer _______________________________________ the information in the reading.

Task 5

The problem is that ____.

According to ______________________, one solution is to ______________________.

Another possibility is to ____.

I think that the best solution is to _____________________ because _________________.

In my opinion, the man/woman should ____.

Task 6

According to the lecturer, ____.

The lecturer states that ____.

The two/major ____________________________ are ______________________________.

The lecturer concludes that ____.

In other words, ____.

Speaking Strategy 14

Use written and spoken contractions

Remember to include contractions in your speech. If you do not use contractions it may sound more like you are reading than speaking. Language that is too informal sounds inappropriate to the raters. They will hear language that does not sound like a serious college student. Natural, spoken English does not include slang, but it does include contractions. Some contractions have written forms with an apostrophe where a letter is deleted. Other contractions do not change their written forms but sounds are deleted when they are spoken.

Written Contractions					***Spoken Contractions***	
I'm	I'll	I've	isn't	I'd	going to	[gonna]
you're	you'll	you've	aren't	you'd	want to	[wanna]
he's	he'll	he's	weren't	he'd	have to	[haveta]
she's	she'll	she's	doesn't	she'd	has to	[hasta]
it's	it'll	it's	don't	it'd	ought to	[oughta]
we're	we'll	we've	hasn't	we'd	like to	[liketa]
they're	they'll	they've	haven't	they'd	should have	[shoulda]
			hadn't		would have	[woulda]
			won't		could have	[coulda]
			shouldn't		must have	[musta]
			wouldn't		might have	[mighta]
			couldn't		got to	[gotta]
					a lot of	[alotta]

CD 2, Track 11

EXAMPLES

1. Formal, written English: If I could meet anyone, I would like to meet Bill Gates.
   Natural, spoken English: If I could meet anyone, I'd <u>like to</u> [liketa] meet Bill Gates.

2. Formal, written English: The university is going to change its policy.
   Natural, spoken English: The university is <u>going to</u> [gonna] change its policy.

3. Formal, written English: In the first place, it is important for teachers to set a good example.
   Natural, spoken English: In the first place, <u>it's</u> important for teachers to set a good example.

4. Formal, written English: The lecturer explained why the experiment did not work.
   Natural, spoken English: The lecturer explained why the experiment <u>didn't</u> work.

5. Formal, written English: I think that the man ought to talk to his professor.
   Natural, spoken English: I think that the man ought to [oughta] talk to his professor.

6. Formal, written English: The lecturer claims that his research does not support the theory.
   Natural, spoken English: The lecturer claims that his research doesn't support the theory.

7. Formal, written English: If visitors came to my hometown, I would take them to the beach.
   Natural, spoken English: If visitors came to my hometown, I'd take them to the beach.

8. Formal, written English: According to the reading passage, glaciers could have caused the unusual topographical features.
   Natural, spoken English: According to the reading passage, glaciers could have [coulda] caused the unusual topographical features.

9. Formal, written English: I admire him because he is a very honest person.
   Natural, spoken English: I admire him because he's a very honest person.

10. Formal, written English: According to the professor, they have formed over millions of years.
    Natural, spoken English: According to the professor, they've formed over millions of years.

CD 2, Track 11 PAUSE

Practice 14

Find the words in the following sentences that could be contractions or spoken forms in natural, spoken English. First rewrite the sentences using contractions. Put the spoken forms in brackets to show that they cannot be written in that way. Then read your sentences aloud. Be sure to record your responses and compare them with the example answers in the Answer Key on page 354.

CD 2, Track 11 RESUME

1. Formal, written English: It is clear that suburbs have been an important part of urban development since the 1920s.

2. Formal, written English: If the university increases the fees, students will have to pay retroactively.

3. Formal, written English: The woman's problem is that she is not eligible for a scholarship because she is an out-of-state student.

4. Formal, written English: The woman suggests that the man register for a class that will not conflict with his lab.

5. Formal, written English: The man's friend thinks that he should have taken the work study position.

6. Formal, written English: I do not agree with the idea for three reasons.

7. Formal, written English: I think that the woman should apply for an internship because she will gain valuable experience.

8. Formal, written English: According to the lecturer, there might have been water on other planets.

9. Formal, written English: In the lecture on satire, the professor states that it has been used in both visual and written works.

10. Formal, written English: A lot of people are participating in the research studies.

You should use formal English in your writing, but be sure to use natural, spoken English on the Speaking Section of the TOEFL.

Speaking Strategy 15

Use colloquial vocabulary

You may hear slang in the conversations in the Listening section and the Speaking section, but your responses to the speaking tasks should not include slang expressions because you will not sound like a serious college student. Prefer colloquial English, which is between formal English and slang.

Slang Expressions	***Colloquial Forms***
For People	
bunch	group, a number
dude	man, friend
fellow	man, friend
guy	man, friend
kid/kids	child, children
buddies	friends
For Things	
junk	things
stuff	things
For Approval	
awesome	impressive, great
cool	good, beautiful, nice
no big deal	not a problem
For Disapproval	
Whatever	If you say so
For Emphasis	
awful	very
real	very, quite
like	[unnecessary word]
Verbs	
freak out	(become) upset
gross out	disgust
hang out	spend time
mess up	make a mistake
be bummed out [bummer]	feel disappointed [disappointment]

EXAMPLES

1. Slang: I like to hang out with my friends.
   Colloquial English: I like to spend time with my friends.

2. Slang: The city was awesome!
   Colloquial English: The city was impressive!

3. Slang: I think he should, like, make an appointment with his advisor, like, right away.
   Colloquial English: I think he should make an appointment with his advisor right away.

4. Slang: I was real lonely when I first arrived on campus.
   Colloquial English: I was very lonely when I first arrived on campus.

5. Slang: The professor explains how kids learn to talk.
   Colloquial English: The professor explains how children learn to talk.

6. Slang: The best birthday I ever had was when a bunch of guys gave me a party.
   Colloquial English: The best birthday I ever had was when a group of friends gave me a party.

7. Slang: According to the professor, it was no big deal.
   Colloquial English: According to the professor, it was not a problem.

8. Slang: I learned a lot of stuff from my English teacher.
   Colloquial English: I learned a lot of things from my English teacher.

9. Slang: The speaker was bummed out about the change in the policy.
   Colloquial English: The speaker was disappointed about the change in the policy.

10. Slang: The coolest gift that I have ever received was a ring from my mother.
    Colloquial English: The nicest gift that I have ever received was a ring from my mother.

Practice 15

Find the slang words in the following sentences. Rewrite those sentences using colloquial English. Then compare your answers with the example answers in the Answer Key on page 355.

1. According to the professor, the study was real successful.

2. The speaker claimed that the adjustment in fees was going to freak out a bunch of students who had not planned for an increase.

3. After class, I usually hang out with some guys from my dorm.

4. The lecturer said that kids who have asthma may experience a spontaneous cure when they grow up because their immune system improves in adulthood.

5. When I go back to my country, I will have to take a lot of stuff for my family and friends.

6. In my opinion, like the best way to prepare for the TOEFL is to take practice tests.

7. I have had several cool teachers, but the best was Mr. Young because he was very knowledgeable and he made learning fun.

8. The man's problem was that he messed up about the date of the field trip and he scheduled two important meetings with his study group in another class.

9. I agree that attending an Ivy League school is totally awesome.

10. A bunch of students signed a petition to change the payment policy for dormitory residents.

Speaking Strategy 16

Learn how to speak from notes

Write down words and a few key phrases to help you remember what you want to say. If you don't feel confident speaking from notes that contain only words, you may want to build up to it in three steps. You can begin by writing out a script. A script includes everything that you plan to say. It is written in complete sentences. Then take the script and find the most important phrases. Finally, use those phrases to find words and a few key phrases for your notes. If you try to write a script during the preparation time, you won't be able to finish it and you won't have the notes you need to support you at the end of your talk. If you write phrases, that won't take as long to complete as a script, but you still may run out of time before the beep that signals you to begin speaking. You need to learn how to speak from notes.

EXAMPLE TASK 2

Some people think that it is important to set goals in order to succeed. Other people think that good luck influences success. Which do you think is more important—goals or luck—and why do you think so? Use specific reasons or examples to support your opinion.

Script

I think it's important to set goals and work hard to achieve them. For one thing, goals help you think clearly about what you're doing. By identifying goals, you get organized. Another thing is the advantage of sharing goals with people who can help you. If others know what you're trying to do, sometimes they can give you advice or assistance. As for good luck, I agree that a lucky opportunity or a chance meeting with a person who can help you is sometimes the key to success...but unless you know where you're going and what you want to achieve, you might not even recognize a lucky opportunity. So, in my opinion, the best way to succeed is to know what you want and work to create your own good fortune. Then, if luck shines on you, you'll just reach your goal sooner.

Phrases

important to set goals
work hard to achieve goals
help you think clearly
you get organized
advantage of sharing goals
people give you advice or assistance
lucky opportunity or chance meeting
a key to success
unless know where you're going
what you want
might not recognize a lucky opportunity
best way to succeed
know what you want
work to create good fortune
if luck shines reach goals sooner

Words

set goals
- work hard
- think clearly

get organized
- advantage sharing
- advice or assistance

lucky opportunity/chance meeting
key to success
- where you're going
- what you want
- might not recognize

to succeed
- know what/where
- work to create

if luck—goal sooner

Practice 16

Use the script to make notes. Try to write words or phrases instead of sentences. Then, using your notes, respond to the question. Be sure to record your response and compare it with the example answer in the Answer Key on page 356.

EXAMPLE TASK 6

Using the main points and examples from the lecture, discuss the Dark Ages and provide examples.

Script

CD 2, Track 12

In the professor's opinion, the term *Dark Ages* is inappropriate and unfair. Modern scholars prefer the term *Early Middle Ages* to refer to the period between 400 and 1,000 C.E. He claims that merchants and crusaders brought culture and knowledge from the East to Europe. For example, Islamic education was influential when the first medieval universities were established in Europe. He also mentions that books were available and literacy rates had gone up from previous times. Art, architecture, and music began to develop along classical lines. Although he admits that the decline of the Roman Empire resulted in wars and political disorganization, he says that the political and economic systems were already reorganizing around a feudal system. He thinks it's unfair to compare the Early Middle Ages to later periods in history like the Enlightenment. Within the historical context, the Early Middle Ages includes many achievements. It's not the *period* that is dark. It is our *lack of information* about it.

Words or Phrases

Speaking Strategy 17

Make connections using extension and contradiction

Decide whether the integrated question asks you for an extension of the information or a contradiction of the ideas. In Task 3, extension means that the speakers agree with information from the announcement in the reading. Contradiction means that the speakers disagree with the announcement. In Task 4, extension means that the concept in the reading is explained by an example in the lecture. Contradiction means that the example casts doubt on the concept.

EXAMPLES

Task 3

The woman expresses her opinion of the campus announcement. State her opinion and the reasons that she has for holding that opinion.

Notes

Summary of announcement
↓
Transition
↓
Woman's opinion
↓
Reason
↓
Reason

Task 4

The professor discusses an example. Explain how the example in the lecture is related to the concept in the reading.

Notes

Summary of concept

↓

Transition

↓

Example

↓

Relationship

Practice 17

First read the passage. Then listen to the conversation or lecture. Next identify the connection—contradiction or extension—and draw an outline that shows how you will answer the question. Write your notes in the outline. Compare your answer with the outline and notes in the Answer Key on page 357. Finally, listen to the example answer as you follow along with your notes.

Task 3

In Question 3 you will read a short passage and listen to a talk on the same topic. Then you will listen for a question about them. After you hear the question, you have 30 seconds to prepare and 60 seconds to record your answer.

Now read a passage from a campus flyer about an opportunity to participate in a research study.

READING TIME: 45 SECONDS

Research Study

The Psychology Department needs fifty participants for a research study on relationships in college. Undergraduate students 18–21 years of age who are enrolled in a full-time program of studies are eligible for the study. Interested participants should report to the Psychology Office on the second floor of Harmon Hall at 2:00 P.M. on Friday, January 29. It should take about one hour to complete the 100-question survey; however, a commitment of an additional half hour will be required in order to confirm qualifications. Please bring your student ID and a driver's license to verify student status and age. A payment of $20 will be issued at the end of the session.

Now listen to two students who are talking about the research study. The man expresses his opinion of the opportunity. Report his opinion and explain the reasons that he has for having that opinion.

CD 2, Track 13

Notes

CD 2, Track 13 PAUSE

Task 4

In Question 4 you will read a short passage and listen to part of a lecture on the same topic. Then you will listen for a question about them. After you hear the question, you have 30 seconds to prepare and 60 seconds to record your answer.

Now read the passage about ergonomics.

READING TIME: 45 SECONDS

> **Ergonomics**
> The term *ergonomics* is derived from *nomos*, which means "laws," and *ergon*, which means "work." It is the scientific discipline that studies ways in which a job or the equipment used in the work environment can be adapted to human workers. Drawing upon various disciplines, including human biology, psychology, and engineering, ergonomics provides an interdisciplinary approach to practical problems in the workplace. The application of research in ergonomics impacts the design of the environment, systems, and objects that human workers use in their jobs. By making adjustments to support the efficient use of human energy, operator fatigue, discomfort, and injury can be reduced while safety and productivity can be increased.

Now listen to part of a lecture in a business class. The professor is talking about ergonomics. Using information from both the reading passage and the lecture, discuss ergonomics and provide examples from the case study.

CD 2, Track 13 RESUME

Outline

★Bonus: iBT Writing

The integrated essay on the iBT Writing section also requires that you make connections between the reading and listening passages.

Speaking Strategy 18

Stop when you have finished answering the question

Your answer should be complete, but you should not thank the listener and you should not ask the listener to excuse you because your English is not as good as you would like it to be. This is appropriate in many cultures, and even in some situations when you will give a speech in English, but you will lose points for appreciation and apologies at the end of talks in the Speaking section. To signal that you are beginning your conclusion, you may use phrases like *in conclusion, finally,* or *in summary*. The more informal phrase *and so* is also acceptable for TOEFL Speaking tasks. A good conclusion is often a short summary or a logical extension of the information in the talk.

Examples of CORRECT Conclusions

Summary

So spring tides occur when the sun and the moon are aligned, but neap tides occur when the sun and the moon are not in alignment.

Logical Extension

In conclusion, the results of this study could have implications for human populations as well.

Examples of INCORRECT Conclusions

Do NOT end your talks in this way:

In conclusion, let me say thank you for listening to me.

And finally, I am sorry that my English is not good.

And I also want to say thank you for your patience.

In summary, I appreciate your listening to my talk.

Also, I apologize for my bad English.

So, okay. Thank you.

Practice 18

First read each conclusion. Next decide whether it is a good way to end a task on the Speaking section. Put a check mark beside the good conclusions. Then compare your answers with the sentences in the Answer Key on page 359.

1. Finally, let me reiterate the qualities that I require in a friend—loyalty, dependability, and compatibility. And that's why I like Sarah so much.

2. In closing, let me say thank you for giving me your time today. I know that you are very busy.

3. In conclusion, all three types of dinosaurs were common, but they probably did not live during the same time period.

4. So that is my opinion, and I hope that you understand my English.

5. And that's why I prefer public transportation—it's cheaper and it decreases the amount of pollution in the environment.

6. That's all. I'm sorry that I can't think of anything else.

7. So the student disagrees with the university's new policy because it will be inconvenient for students and expensive in the long run.

8. Okay then. We see from the lecture that quilts were a good example of American folk art.

9. So I think that the man should talk with his professor before he makes a decision, because, if the professor says "no," then he can always try the other option.

10. I don't have time to say more. Thank you.

★★**Bonus:** iBT Writing, ITP Essay
A strong conclusion is important for the essays in the iBT and the ITP as well as for all six tasks in the Speaking section of the iBT.

Speaking Strategy 19

Think like a rater

Use the Evaluator's Checklist to estimate your score. Raters are trained to ignore minor errors that they may hear in a test because they do not expect non-native speakers of English to produce perfect speech. In general, if an error interferes with the rater's comprehension, the rater will judge it a major error. If an error does not cause the rater to stop and think in order to interpret the meaning, the rater will judge it a minor error. Major errors are distracting, but minor errors do not confuse or interfere with meaning. Speakers lose points for major errors.

Examples of Evaluator's Checklists

3.5–4 Excellent

✔ The speaker responds to the question.
✔ The content is accurate.
✔ The pace is continuous without constant hesitations.
✔ The speaker's point of view is clear.
✔ The talk is direct and well-organized.
✔ The sentences are logically connected.
✔ Details and examples support the opinion.
✔ The speaker expresses complete thoughts.
✔ The meaning is easy to comprehend.
✔ A wide range of appropriate vocabulary is used.
✔ Errors in grammar are minor or repeated.
✔ The talk is within a range of 125–150 words.

2.5–3 Good

- ✔ The speaker responds to the question.
- ✔ The content is accurate.
- ✔ The pace is continuous with some hesitations.
- ✔ The speaker's point of view is mostly clear.
- ✔ The talk demonstrates basic organizational strategies.
- ✔ The sentences are logically connected.
- ✔ Details and examples support the opinion.
- ✔ The speaker expresses complete thoughts.
- ✔ The meaning requires some effort to comprehend.
- ✔ A basic range of appropriate vocabulary is used.
- ✔ Errors in grammar are repeated.
- ✔ The talk is within a range of 125–50 words.

1.5–2 Limited

- ✔ The speaker includes some basic ideas related to the question.
- ✔ The content is off topic.
- ✔ The pace includes hesitations and long pauses.
- ✔ The speaker's point of view regarding the topic is not clear.
- ✔ The talk is not very well-organized.
- ✔ The sentences include some logical connection.
- ✔ Few details and examples support the opinion.
- ✔ The speaker expresses limited thoughts.
- ✔ The meaning requires effort to comprehend.
- ✔ A minimal range of vocabulary is used.
- ✔ Errors in grammar occur throughout.
- ✔ The talk is not within a range of 125–150 words.

Practice 19

First listen to the lecture and the question. Next listen to three responses and refer to the Checklists to evaluate the speakers. Then compare your evaluations with the evaluations in the Answer Key on page 359.

CD 2, Track 14

Response 1
Level:
Comments:

Response 2
Level:
Comments:

Speaking Strategy 20

Listen to good models

To have the greatest success, you should listen to good models. The models should be like the kind of speaking that you want to do. In order to improve your responses on the Speaking section of the TOEFL, you should listen to good responses to the tasks on the Speaking section of the TOEFL. A number of research studies confirm that listening is one of the best ways to improve speaking.

CD 2, Track 15

EXAMPLES

Listen to the example answers for each task in the Speaking section. Do not look at the scripts the first time that you listen.

Task 1

In your opinion, which invention was the most important contribution to the world? Include specific reasons and details to explain your choice.

Task 2

Some people would like to go to a sports event like a baseball game or a race. Other people would enjoy a cultural activity like a concert or a play. What do you prefer to do, and why? Use specific reasons and examples to support your choice.

Task 3

The man expresses his opinion of the tutoring opportunity. Report his opinion and explain the reasons that he has for having that opinion.

Task 4

Using information from both the reading passage and the lecture, discuss motivation and provide examples.

Task 5

Describe the woman's problem and the solutions that her friend suggests. What do you think the woman should do, and why?

Task 6

Using the main points and examples from the lecture, discuss nanotechnology and provide examples.

Practice 20

Listen again as you read the scripts for the example answers located in the Answer Key. Next, identify how the speaker is using the strategies that you have previously studied in this chapter. Try to identify three strategies for each task. Then compare your answers with those in the Answer Key on page 361.

CD 2, Track 15 REPEAT

Task 1

Task 2

Task 3

Task 4

Task 5

Task 6

5

STRATEGIES FOR THE STRUCTURE SECTION

OVERVIEW OF THE ITP STRUCTURE AND WRITTEN EXPRESSION SECTION

The Structure and Written Expression section is Section 2 on the ITP TOEFL. The Structure and Written Expression section tests your ability to recognize language that is appropriate in standard written English. During the test, you will be asked to find structures that are used correctly and written expressions that are used incorrectly.

In Part A, you will see 15 incomplete sentences. Beneath each sentence, you will see four words or phrases. You are asked to choose the word or phrase that would complete the sentence correctly. Then, on your answer sheet, you fill in the oval that corresponds to the letter of the answer you have chosen.

In Part B, you will see 25 sentences with four underlined words or phrases. You are asked to choose the word or phrase that must be changed in order for the sentence to be correct. Then, on your answer sheet, you fill in the oval that corresponds to the letter of the answer you have chosen.

A Structure and Written Expression section is NOT included in the Internet-Based TOEFL, but the raters will include grammar as part of the evaluation for your Speaking and Writing sections.

You will have 25 minutes to read and answer all of the questions.

The structures in this chapter were chosen because they are commonly tested on the ITP and they are useful in the Speaking and Writing sections of the iBT.

TIPS FROM SUCCESSFUL TOEFL STUDENTS

WHAT TO DO ON THE STRUCTURE SECTION

- Read all of the answer choices before you mark your choice. Sometimes an answer is a good, but not the best, answer. You won't know that until you have considered all four choices.
- To complete the Structure section on time, look at the time after you have answered all of the questions for Part A. You should have been working for 10 minutes or less.
- On the Paper-Based TOEFL, be sure to erase all corrections completely. Sometimes the machine will score your answer sheet incorrectly if there is a stray pencil mark!
- Answer every question. Don't leave a blank. You do not lose points for incorrect answers!
- If you finish the section before time is called, do not go to the next section. If the administrator sees you working on another section, you could be reported, and your test may not be scored.

Structure Strategy 1

Look for the main verb in every sentence

With the exception of commands in which the subject (you) is understood, every English sentence must have a subject and a main verb. The wrong answers will present an *-ing* form, an infinitive, an auxiliary verb only, or another part of speech instead of a main verb.

EXAMPLES

Arizona _______ a very dry climate.

● has

Ⓑ being

Ⓒ to have

Ⓓ with

Only Choice A provides a main verb for the subject *Arizona*. Choice B is not correct because it is an *-ing* form, not a main verb. Choice C is not correct because it is an infinitive. Choice D is not correct because it is a preposition, not a main verb.

Venomous snakes <u>with</u> modified teeth connected to <u>poison glands</u> <u>in which</u> the venom
(A: with; B: poison glands; C: in which)

<u>is secreted</u> and stored.
(D: is secreted)

Choice A is not correct because the sentence does not have a main verb. The word *with* should be *have* to provide a main verb for the subject *snakes*.

Practice 1

First, read the sentence. If it is correct, put a check mark beside it. ✔
If it is incorrect, edit the sentence.
Then check your answers with those in the Answer Key on page 364.

1. Hermit crabs carrying the shells of other species for protection from predators.

2. Based on current predictions, the Earth's population will more than nine billion people by 2040.

3. Whereas planets are formed as a result of natural star formation processes, stars emerging from large clouds of gas and dust.

4. Rice is a labor-intensive plant that grows in flooded fields and thrives in a hot, humid climate.

5. Koalas, known for sleeping during the day, actively feed on Eucalyptus leaves at night.

★★★**Bonus:** iBT Speaking, Writing, ITP Essay
You need to use main verbs correctly in your sentences in both the Speaking section and the Writing sections.

Structure Strategy 2

Match each auxiliary verb with the ending on the main verb

Some English verbs have an auxiliary verb and a main verb. The auxiliary verb contributes the time/tense meaning and the main verb contributes the vocabulary meaning. The most common auxiliary verbs are BE, HAVE, DO, and the MODALS *may*, *might*, *can*, *could*, *will*, *would*, *shall*, *should*, and *must*. Auxiliary verbs are used with main verbs.

BE + -ing	***HAVE + participle***	***MODAL + verb word***
is studying	has studied	will study

EXAMPLES

The giraffe survives in part because it_______the vegetation in the high branches of trees where other animals have not grazed.

Ⓐ to reach

● can reach

Ⓒ reaching

Ⓓ is reach

Only Choice B provides an auxiliary verb that can be used correctly with a main verb. Choice A is not correct because it is an infinitive which cannot be used as a main verb with the subject *giraffe*. Choice C is not correct because it is an *-ing* form without the auxiliary BE. Choice D is not correct because it is an auxiliary BE without the *-ing* form of the main verb.

<u>According</u> to some scientists, the Earth <u>losing</u> <u>its</u> outer atmosphere <u>because of</u> pollutants.
A B C D

Choice B is not correct because it is an *-ing* form without the auxiliary BE. The word *losing* should be *is losing* to provide an auxiliary BE form with the *-ing* form of the main verb.

Practice 2

First, read the sentence. If it is correct, put a check mark beside it. ✔
If it is incorrect, edit the sentence.
Then check your answers with those in the Answer Key on page 364.

1. Because blood from different individuals will have different types of antigens on the surface of the red cells and antibodies in the plasma, a dangerous reaction can to occur between a donor and a recipient in a blood transfusion.

2. In the tundra, beneath a rocky layer of topsoil, a layer of permafrost has remained permanently frozen so that the soil can never get soft or warm enough to cultivate plants.

3. An artist who is created a fresco must have a very clear design prepared in advance because the plaster tends to dry fairly quickly and it is not possible to paint over it with the same ease as it is in an oil painting.

4. The recognized Native American tribes in the United States have form their own governments on reservation lands.

5. Some herds have been returning to the same watering holes for decades, encouraging the belief that elephants never forget.

★★★**Bonus:** iBT Speaking, Writing, ITP Essay
Using auxiliary verbs correctly in both the Speaking section and the Writing sections will add points to your scores on these sections.

Structure Strategy 3

Look for *that* to mark a dependent clause

A main clause, also called an independent clause, can function as a separate sentence. A subordinate clause, also called a dependent clause, must be attached to a main clause. A dependent clause is often marked (preceded) by the clause marker *that*. Look for sentences that use the clause marker incorrectly. The clause marker *that* with a dependent clause is not a sentence. The clause marker *that* with a sentence and no dependent clause following it is not correct. Both main clauses and dependent clauses must have their own verbs.

EXAMPLES

Most beekeepers have observed _______ at the approach of a thunderstorm.

Ⓐ enraging the bees

● that bees become enraged

Ⓒ that bees enraging

Ⓓ the bees to enrage

Only Choice B has a clause marker *that* followed by a clause with a subject *bees* and verb *become enraged*. Choices A, C, and D are not correct because they are missing a clause marker and/or a verb. . . . *that the atmospheric changes enrage the bees* would also be correct.

<u>Thunder</u> <u>that</u> we recorded as audible from distances as far away <u>as</u> ten <u>miles</u>.
A B C D

Choice B is not correct because *that* marks a dependent clause, but there is no main clause in the sentence. *Our instruments identified thunder that we recorded as audible from distances as far away as ten miles* would be correct because it includes a main clause.

Practice 3

First, read the sentence. If it is correct, put a check mark beside it. ✔
If it is incorrect, edit the sentence.
Then check your answers with those in the Answer Key on page 365.

1. Since humans do not inhabit most of the islands in the Galapagos chain, that the animals are not afraid of people.

2. Most democracies employ a system of checks and balances to ensure that none of the branches of government too powerful.

3. The Homestead Act provided 160 acres to pioneers who remained on and improved the land in Western territories of the United States.

4. A researcher at the University of Toronto has developed a contact lens can monitor blood sugar levels in diabetics.

5. Some astronomers claim that more than 90 percent of the mass of the universe is dark matter which can't be seen.

★★**Bonus:** iBT Writing, ITP Essay
Understanding sentences and clauses will help you when you edit your essays on the Writing sections.

Structure Strategy 4

Check every sentence for subject–verb agreement

Agreement means that the subject and verb have the same person and number (singular or plural). The wrong answers will often present a verb that agrees with the modifier of a subject instead of the subject itself.

EXAMPLES

Groups of tissues, each with its own function, _______ in the human body.

Ⓐ it makes up the organs

● make up the organs

Ⓒ they make up the organs

Ⓓ makes up the organs

Only Choice B provides a verb that agrees with the plural subject *Groups of tissues*. Choice A is not correct because *it* is a redundant subject (repetition of the subject) and the verb *makes* is singular. Choice C is not correct because *they* is a redundant subject. Choice D is not correct because the verb *makes* is a singular, not a plural verb.

The Zoning Improvement Plan, better known as Zip Codes (A), enable (B) postal clerks to speed (C) the routing of an ever-increasing volume of mail (D).

Choice B is not correct because the plural verb does not agree with the singular subject *The Zoning Improvement Plan*. The word *Codes* in the modifier is confusing, but the verb must agree with the original subject, not the modifier. *Codes enable postal clerks to speed the mail* would also be correct.

Practice 4

First, read the sentence. If it is correct, put a check mark beside it. ✔
If it is incorrect, edit the sentence.
Then check your answers with those in the Answer Key on page 365.

1. Columbus, one of the state's largest cities, are not only the capital of Ohio but also a typical metropolitan area, often used in market research.

2. The cerebrum, which is comprised of both right and left hemispheres, control the muscles of the opposite side of the body.

3. Influenced by the Phoenician system, the Greek alphabet, which includes 24 letters, were developed in 1,000 B.C.E.

4. Because some varieties of bacteria produce hydrogen, they are being considered as a potential energy source.

5. Flutes carved from bone are among the earliest musical instruments found in archaeological sites.

★★**Bonus:** iBT Writing, ITP Essay
Reading your essays to check for subject–verb agreement is one of the important skills for editing the Writing sections.

Structure Strategy 5

Match complements with the verbs that precede them

Some verbs require an infinitive and other verbs require an *-ing* form after them in the complement.

Infinitive			*-ing Form*	
agree	fail	plan	admit	miss
appear	forget	prepare	appreciate	postpone
arrange	hesitate	pretend	avoid	practice
ask	hope	promise	complete	quit
claim	intend	refuse	consider	recall
consent	learn	seem	delay	recommend
decide	manage	tend	deny	regret
demand	mean	threaten	discuss	risk
deserve	need	wait	enjoy	stop
expect	offer	want	finish	suggest
			keep	tolerate
			mention	understand

EXAMPLES

Strauss finished _______ two of his published compositions before his tenth birthday.

Ⓐ written

Ⓑ write

Ⓒ to write

● writing

Only the *-ing* form in Choice D is used in the complement after the verb *finished*.

Representative democracy seemed <u>evolve</u> <u>simultaneously</u> <u>during</u> the eighteenth and
(A: evolve; B: simultaneously; C: during)

nineteenth centuries in Britain, Europe, and <u>the</u> United States.
(D: the)

Choice A is not correct because an infinitive should be used in the complement after the verb *seemed*.

Practice 5

First, read the sentence. If it is correct, put a check mark beside it. ✔
If it is incorrect, edit the sentence.
Then check your answers with those in the Answer Key on page 366.

1. Constellations tend to change over time because every star is constantly moving through space and the pattern disappears.

2. Repression occurs when an individual avoids to think about painful experiences, and can even temporarily forget about them.

3. In some optical illusions, an image seems remaining after the exposure to the original image has ended.

4. Most doctors agree that when patients stop to smoke, their lungs repair themselves within seven years.

5. The phoenix is a mythical bird that appears to emerge from the ashes of its funeral pyre.

★★★**Bonus:** iBT Speaking, Writing, ITP Essay
Using the correct complement after common verbs will improve both your speech and your writing on the Speaking section and the Writing section.

Structure Strategy 6

Locate passive word order to emphasize process or result

In a passive sentence, the actor is unknown or not important. The subject is usually not a person. The word order for a passive is BE + participle. Passive sentences are especially common in certain styles of scientific and academic writing when the focus is on the process or result.

Examples

In stringed instruments, tones _______ by playing a bow across a set of strings that may be made of wire or gut.

Ⓐ they produce

Ⓑ producing

● are produced

Ⓓ that are producing

Only Choice C provides a passive word order that gives *the tones* more importance than the person playing the tones. None of the other choices are passives.

<u>Work</u> <u>is</u> often <u>measure</u> in units <u>called</u> foot pounds.
A B C D

Choice C is incorrect because in this case, the passive requires a participle. The word *measure* should be *measured*.

Practice 6

First, read the sentence. If it is correct, put a check mark beside it. ✔
If it is incorrect, edit the sentence.
Then check your answers with those in the Answer Key on page 366.

1. Cupid, one of the ancient Roman gods, represented as a little child with wings.

2. Our solar system is divide into three categories of planets based on their size and density.

3. Although a baby is born with 300 bones, several will fuse together by adulthood when only a few more than 200 will normally be accounted for.

4. More than forty countries found in Asia, which is the world's largest continent.

5. The first bank-issued credit cards were offered by the Flatbush National Bank of Brooklyn, New York.

★★★★**Bonus:** iBT Reading, Speaking, Writing, ITP Essay
Passive sentences will often occur in the academic reading passages on the Reading, Speaking, and Writing sections. Passive sentences are also typical of good academic essays.

Structure Strategy 7

Expect a verb word after subjunctive verbs and expressions

Remember that a *verb word* is the infinitive without the word *to*. A special group of verbs and expressions are used before *that* and a verb word clause to express possibilities and importance.

Verbs

ask	insist	request
advise	prefer	require
demand	propose	suggest
desire	recommend	urge

Expressions

It is crucial	It is important
It is essential	It is necessary
It is imperative	It is vital

EXAMPLES

Many architects prefer that a dome _______ used to roof buildings in which floor space needs to be conserved.

Ⓐ will

Ⓑ are

Ⓒ being

● be

Only Choice D provides a verb word for the clause following *prefer*. Choice A is not correct because it is a modal, not a verb word. Choice B is not correct because it is a present tense verb. Choice C is not correct because it is an *-ing* form.

It is essential that vitamins <u>are</u> (A) supplied either by food <u>or</u> (B) by <u>supplements</u> (C) for normal growth <u>to occur</u> (D).

Choice A is not correct because a verb word must be used in the clause after the expression *it is essential*. The word *are* should be *be*.

Practice 7

First, read the sentence. If it is correct, put a check mark beside it. ✔
If it is incorrect, edit the sentence.
Then check your answers with those in the Answer Key on page 366.

1. It is essential that antioxidants, which are absorbed primarily from fruit and vegetables, be present to repair cells.

2. It is important that renewable energy resources are explored in order to provide alternatives to fossil fuels.

3. In 1773, after the Americans demanded that the British repealed the tax on tea, their refusal instigated the famous Boston Tea Party.

4. It is vital to the pollination of many crops that researchers can discover why colonies of honeybees are dying.

5. According to studies, infants prefer that an object have figures and patterns on it rather than solid colors.

★★★**Bonus:** iBT Speaking, Writing, ITP Essay
If you use subjunctive verbs and expressions correctly to emphasize importance in your sentences on both the Speaking section and the Writing sections, it will make a positive impression on the raters.

Structure Strategy 8

Recognize conditionals that express scientific facts

Absolute conditionals express scientific facts. After the condition, *will* and a verb word confirm that the result is absolutely certain. A present tense verb is also correct.

EXAMPLES

If water is heated to 212 degrees F _______ as steam.

- ⬤ it will boil and escape
- Ⓑ it is boiling and escaping
- Ⓒ it boil and escape
- Ⓓ it would boil and escape

Only Choice A provides a clause of result with *will* and a verb word. Choice B is not correct because BE with an *-ing* form is in the clause of result. Choice C is not correct because *will* is missing before the verb word. Choice D is not correct because *would* is used instead of *will*.

If a live sponge <u>is broken</u> (A) into pieces, each piece <u>would turn</u> (B) into a new sponge <u>like</u> (C) <u>the original one</u> (D).

Choice B is not correct because *will* must be used in the clause of result. The word *would* should be *will*. The present tense *turns* is also correct.

Practice 8

First, read the sentence. If it is correct, put a check mark beside it. ✔
If it is incorrect, edit the sentence.
Then check your answers with those in the Answer Key on page 367.

1. If children are healthy, they are learning to walk at about eighteen months old.

2. If the trajectory of a satellite will be off at launch, it will get worse as the flight progresses.

3. If light strikes a rough surface, it diffused because the rays are not parallel to each other.

4. If the President ignores a bill while Congress is in session, it will become a law after ten days, even without a signature.

5. If humans are deprived of sleep, the immune system weakens.

★★★**Bonus:** iBT Reading, Speaking, Writing
Absolute conditionals are used in the writing found in scientific passages on the Reading, Speaking, and Writing sections. It is important to understand them.

Structure Strategy 9

Identify subjects beginning with *that*

The subject of a verb can be a single noun, a long noun phrase, or a long noun clause. Like all clauses, the nominal *that* clause has a subject and verb. The nominal *that* clause functions as the subject of the main verb.

Subject	**Verb**
Single Noun	
That universities require standardized test scores	is usual.
Long noun phrase	
That more competitive colleges and universities require standardized test scores	is usual.
Long noun clause	
That universities that have more competitive criteria for admission require standardized test scores	is usual.

EXAMPLES

_______ migrate long distances is well documented.

Ⓐ That it is birds

● That birds

Ⓒ Birds that

Ⓓ It is that birds

Only Choice B is the subject of the verb *migrate* in the clause that functions as a subject of the main verb *is*. Choices A and D are not correct because they include the additional verb *is*. Choice C is not correct because *Birds* is a plural subject which does not agree with the main verb *is*.

That <u>it is</u> the moon influences only <u>one kind</u> of tide is not <u>generally</u> <u>known.</u>
A B C D

Choice A is incorrect because *That the moon* is the subject of the main verb *is known*. The subject and verb *it is* are redundant and should be deleted.

Practice 9

First, read the sentence. If it is correct, put a check mark beside it. ✔
If it is incorrect, edit the sentence.
Then check your answers with those in the Answer Key on page 367.

1. Comets have periodic orbits was confirmed in 1758 by the appearance of a comet when Halley predicted it.

2. That Michelangelo was highly regarded by his contemporaries is evidenced by the fact that two biographies were published during his lifetime.

3. That the discovery that about 8 percent of the Earth's crust is made of aluminum caused the price to fall from $20 to $1 in the late 1800s.

4. That seventy-six million American children were born between 1945 and 1964 is significant because the demographic affects marketing trends.

5. Shakespeare's plays that they were written by another author has not been supported by Elizabethan documents.

★★★**Bonus:** iBT Speaking, Writing, ITP Essay
Using a variety of sentences on both the Speaking section and the Writing sections makes your speaking and essays more interesting. Nominal *that* clauses are complex. When you use one correctly in a task, it is memorable for raters. Just don't use them all the time—it diminishes the effect.

Structure Strategy 10

Watch for clauses that begin with *it*

An anticipatory *it* clause expresses belief or knowledge. Anticipatory means *before*. Some *it* clauses that go before main clauses to express belief are listed below. Look for a main subject and a main verb after the anticipatory *it* clause.

It is accepted	It is supposed
It is assumed	It is said
It is believed	It is stated
It is hypothesized	It is thought
It is known	It is true
It is proposed	It is written

EXAMPLES

_______ the Giant Ape Man, our biggest and probably one of our first humanoid ancestors, was just about the size of a male gorilla.

● It is believed that

Ⓑ That it is

Ⓒ That is believed

Ⓓ That believing

Only Choice A is an anticipatory *it* clause that introduces the main clause *Giant Ape Man . . . was*. Choices B, C, and D are not correct because they confuse the anticipatory *it* clause with a nominal *that* clause. *That the Giant Ape Man...was just about the size of a male gorilla is generally accepted* would also be correct.

<u>That it is believed</u> (A) that <u>most of the earthquakes</u> (B) in the world occur <u>near</u> (C) <u>the youngest</u> (D) mountain ranges—the Himalayas, the Andes, and the Sierra Nevadas.

Choice A is not correct because the anticipatory *it* clause does not include the word *That*. The anticipatory *it* clause introduces the main subject *most of the earthquakes* and the main verb *occur*.

Practice 10

First, read the sentence. If it is correct, put a check mark beside it. ✔
If it is incorrect, edit the sentence.
Then check your answers with those in the Answer Key on page 367.

1. It believed that no two tigers have exactly the same pattern of stripes.

2. That it is proposed that the sun accounts for 99 percent of the matter in the solar system.

3. It is accepted that maps distort the size and shape of the geographic features because they are flat.

4. It is well known that the best quality charcoal for sketching pencils is made from vine wood heated in a kiln until only the carbon remains.

5. That is thought that the Moon may have had life on it at one time because there is evidence of water at the bottom of the Cabeus Crater.

★★★**Bonus:** iBT Speaking, Writing, ITP Essay
The anticipatory *it* is another way to introduce variation in your speaking and writing. Used occasionally, this structure demonstrates a high level of skill to raters on the Speaking and Writing sections.

Structure Strategy 11

Look for time phrases after *for and since*

Both *for* and *since* express a duration of time. *For* is used before a quantity of time and *since* is used before a specific time. *For* answers the question: *how long*? For example, *for a month*, *for two years*. *Since* answers the question: *beginning when*? For example, *since Wednesday*, *since July*, *since 2010*. The verb structure HAVE and a participle is often used with duration. *Since . . . ago* is an unusual but correct form that includes both *beginning when* and *how long*. It is followed by a specific plural time. For example, *since* two years *ago*.

EXAMPLES

Penguins, the most highly specialized of all aquatic birds, may live _______ twenty years.

Ⓐ before

Ⓑ since

● for

Ⓓ from

Only Choice C is used before the quantity of time *twenty years*. Also correct: Penguins in the area have been studied *since* 1960. Penguins in this area have been studied *since* fifty years *ago.*

Because national statistics on crime have only been kept <u>for 1930</u>, <u>it is</u> not possible
(A: for 1930; B: it is)

<u>to make</u> judgments about crime <u>during the early years</u> of the nation.
(C: to make; D: during the early years)

Choice A is not correct because *for* should be followed by a quantity of time, not a specific time. For example, *since 1930* or *for 80 years* would be correct.

Practice 11

First, read the sentence. If it is correct, put a check mark beside it. ✔
If it is incorrect, edit the sentence.
Then check your answers with those in the Answer Key on page 368.

1. Found in a wide variety of habitats, some snails have been known to hibernate for three years.

2. Hadrian's Wall has survived since 2,000 years, a tribute to the architects and engineers as well as to the Roman armies that built it.

3. Forest biomes have been evolving since about 420 million years ago.

4. Although it influenced other styles of art, Impressionism in its purest form lasted since only about 15 years.

5. New Zealand was the first country to grant women the vote, a right which they have enjoyed for 1893, almost 40 years before women were granted the vote in the United States.

★★★**Bonus:** iBT Speaking, Writing, ITP Essay
This is a minor error in speech and writing that does not interfere with communication and is relatively easy to correct. When you use *for* and *since* correctly on the Speaking and Writing sections, you do not interrupt the flow of language with this minor error.

Structure Strategy 12

Check nouns in noncount categories

Noncount nouns have only one form. They are used in agreement with singular verbs. Usually the word *the* does not precede a noncount noun unless it refers to a specific example. The following categories of noncount nouns are listed for your reference:

1. Food staples purchased in various forms: *bread, meat, butter*
2. Construction materials that can change shape, depending on the product: *wood, iron, glass*
3. Liquids that can change shape, depending on the container: *oil, tea, milk*
4. Natural substances that can change shape, depending on natural laws: *steam, water, ice*
5. Substances that have many small parts: *rice, sand, sugar*
6. Groups of similar items: *clothing* (a coat, a shirt, a sock), *furniture* (a table, a chair, a bed)
7. Languages: *Arabic, Japanese, Spanish*
8. Abstract concepts, often with endings *-ness*, *-ance*, *-ence*, *-ity*: *beauty, ignorance, peace*
9. Most *-ing* forms: *learning, shopping, working*
10. Activities: chess, soccer, tennis

EXAMPLES

_______ at 212 degrees F and freezes at 32 degrees F.

Ⓐ Waters boils

Ⓑ The water boils

● Water boils

Ⓓ Waters boil

Only Choice C provides a noncount noun with a singular verb. Choices A and D are not correct because *water* is a noncount noun, not a plural noun, when it is used to refer to a natural substance that can change shape according to natural laws. Choice B is not correct because the word *the* should not be used with a noncount noun.

<u>The iron</u> (A) was used in its native state as an iron-nickel alloy thousands of years before <u>the Iron Age</u> (B) because it was not necessary for <u>prehistoric man</u> (C) to smelt it in order to make <u>tools</u> (D).

Choice A is not correct because it refers to construction material that can change shape depending on what is made. It is a noncount noun that should not be preceded by *the*. Choice B,*The Iron Age,* is not a noncount noun because it refers to a specific time period. Choice C is a noncount noun because it refers to all prehistoric humans. Choice D is a noncount noun because it refers to a group of things in different sizes and shapes. *The iron that was found in the archaeological site* . . . would also be correct because it refers to a specific example of iron—*the* iron *in the site*.

Practice 12

First, read the sentence. If it is correct, put a check mark beside it. ✔
If it is incorrect, edit the sentence.
Then check your answers with those in the Answer Key on page 368.

1. About 98 percent of Antarctica is covered by the ice, extending over an area of fourteen million square kilometers.

2. Although hundreds of languages are recognized in India, Hindi is considered the primary official language and the English is recognized as a secondary official language.

3. For thousands of years, wood has been used for both fuel and construction material in areas where trees are plentiful.

4. Flooding the fields for three days or less is a traditional method for controlling weeds in many parts of the world where rice is cultivated.

5. Evidence of the furniture survives from the Neolithic period, including stone beds, chairs, tables, and dressers.

★★★**Bonus:** iBT Speaking, Writing, ITP Essay
By identifying noncount nouns, you will also be able to select the correct articles, adjectives, and verbs to use on both the Speaking section and the Writing sections.

Structure Strategy 13

Find noncount nouns in English that are count nouns in other languages

Nouns are especially troublesome when they are classified differently in your language. For your reference, some of the most common have been listed for you below.

advice	ignorance	news
anger	information	patience
courage	knowledge	permission
damage	leisure	poetry
equipment	luck	poverty
fun	money	progress
homework	music	work

EXAMPLES

Fire-resistant materials are used to retard _______ of aircraft in case of accidents.

Ⓐ a damage to the passenger cabin

Ⓑ that damages the passenger cabin

● damage to the passenger cabin

Ⓓ passenger cabin's damages

Only Choice C provides a noncount noun. Choice A is not correct because *damage* is a noncount noun, not a singular noun. The article *a* is not used with a noncount noun. Choice B is not correct because *damages* is used as a verb without a subject. The clause *fire that damages the passenger cabin* would also be correct. Choice D is not correct because *damage* is a noncount noun, not a plural noun.

A progress has been made toward finding a cure for many types of cancer.
(A: A progress; B: has been made; C: finding; D: cancer)

Choice A is not correct because *progress* is a noncount noun that cannot be preceded by the article *a*. Choice D is correct because it is a noncount noun that includes groups of similar items, in this case, different types of cancer. *A lot of progress* would also be correct.

Practice 13

First, read the sentence. If it is correct, put a check mark beside it. ✔
If it is incorrect, edit the sentence.
Then check your answers with those in the Answer Key on page 368.

1. The Cabinet and the White House staff provide advice to the President.

2. Because so many mobile devices are available at relatively low cost, informations from international sources is available instantaneously.

3. College students should calculate two hours of homeworks for every hour of class time that they schedule.

4. From space, astronomers use special telescopes with cameras and other equipments that allow them to study X-ray and gamma ray emissions.

5. Works of music cannot be copied without permission because an artist's work is protected by law.

★★★**Bonus:** iBT Speaking, Writing, ITP Essay
By using this group of noncount nouns correctly, you will improve your Speaking section and Writing sections.

Structure Strategy 14

Notice infinitives and *-ing* forms that begin sentences and clauses

Remember that infinitives and *-ing* forms can be used as subjects of sentences and clauses and they agree with singular verbs. Verb words (the infinitive without *to*) or *to* with an *-ing* form are incorrect options that are presented to confuse you.

EXAMPLES

_______ trees is a custom that many people engage in to celebrate Arbor Day.

Ⓐ The plant

Ⓑ Plant

● Planting

Ⓓ To planting

Only the *-ing* form in Choice C is used as the subject of the verb *is*. The infinitive *to plant* would also be correct.

With the aid of <u>voice recognition programs</u> (A), <u>dictate</u> (B) directly onto a computer has become an easier option <u>than</u> (C) <u>typing</u> (D) for many people.

Choice B is not correct because either the infinitive *to dictate* or the *-ing* form *dictating* should be used as the subject of the verb *has become*.

Practice 14

First, read the sentence. If it is correct, put a check mark beside it. ✔
If it is incorrect, edit the sentence.
Then check your answers with those in the Answer Key on page 369.

1. Carve, using either wood or stone, is one of the oldest and most respected crafts worldwide.

2. Recent studies suggest that sleeping for one hour during the day may not disturb the nightly routine as previously thought.

3. To advertising has changed in response to the increased popularity of Internet sites and mobile devices.

4. According to paleontologists, scientists who study fossils, the classifying dinosaurs can be complicated because they demonstrate a very large range of characteristics.

5. Although using solar energy does not produce air pollution and water pollution, the materials and chemicals employed in the manufacturing process of the cells do impact the environment indirectly.

★★★**Bonus:** iBT Speaking, Writing, ITP Essay
Starting with an infinitive or *-ing* subject is another opportunity to use a variety of sentence structures, which will help you achieve a higher score on the Speaking section and the Writing sections.

Structure Strategy 15

Distinguish between cause and result adjectives

An *-ing* form that functions as an adjective usually expresses cause. It is derived from an active verb. An *-ed* adjective usually expresses result. It is derived from a passive verb.

EXAMPLES

The *Canterbury Tales*, written about 1386, is as alive and _______ today as it was hundreds of years ago.

Ⓐ appealed

Ⓑ appeal

● appealing

Ⓓ the appeal of

Only Choice C provides an adjective that describes the book as *causing* a positive result. Choice A is not correct because it is the reader, not the *Canterbury Tales*, to whom the *Tales* appealed. Choice B is not correct because it is a verb word, not an adjective. Choice D is not correct because it is a noun phrase, not an adjective.

The Arabs, who (A) possessed (B) a remarkable gift for astronomy, mathematics, and geometry, were (C) also known for making challenged (D) but accurate maps for explorers.

Choice D is not correct because the maps are *causing* a result. *Maps* cause the *explorers* to be *challenged* as a result of using them, but the maps are *challenging*.

Practice 15

First, read the sentence. If it is correct, put a check mark beside it. ✔
If it is incorrect, edit the sentence.
Then check your answers with those in the Answer Key on page 369.

1. Archipelagos, a chain or cluster of islands, tend to form in isolating parts of the ocean.

2. After years of watching silent films, in 1927 audiences saw *The Jazz Singer*, an astonished full-length feature with dialogue and singing.

3. Light pollution from bright, artificial lights can affect ecosystems because it can cause insects and birds to become disoriented.

4. Some of the most challenged mountains to climb in North America can be found in the Canadian provinces of British Columbia and Alberta.

5. The origin of the word *Google* is interesting because it was inspired by *googol*, the number 1 with one hundred zeros after it.

★★★**Bonus:** iBT Speaking, Writing, ITP Essay
Using correct grammar for minor structures like *-ed* and *-ing* adjectives can make a difference between a good score and a great score on both the Speaking section and the Writing sections.

Structure Strategy 16

Double check the small words in comparative estimates

Comparative phrases can express estimates. Some phrases are single comparisons, used before a noun. For example, you have read *almost all of the books*. Other phrases are double comparisons, used to show the relationship between two events. For example, *the more* you review, *the better* you will score. Still other phrases are multiple comparisons, used with multiple numbers like *twice, three times*, *four times*, etc. For example, you probably know *three times as much as* you did when you started!

Single comparisons	***Double comparisons***		***Multiple comparisons***
almost all of the	the more	the less	twice as much as
most of the	the —	the —	— as much as
			— as many as

EXAMPLES

It is generally true that the lower the stock market falls, _______.

Ⓐ higher the price of gold rises

Ⓑ the price of gold rises high

● the higher the price of gold rises

Ⓓ rises high the price of gold

Only Choice C provides the second part of a double comparison. Both clauses contain *the* with a comparative, and they influence each other. Choice A is not correct because it is missing the word *the* at the beginning of the comparison. Choices B and D are not correct because they do not include comparatives.

After the purchase of the Louisiana Territory, the United States had <u>twice more</u> (A) land as <u>it</u> (B) had previously owned, including <u>more than</u> (C) two million square kilometers <u>which</u> (D) extended from the Mississippi River to the Rocky Mountains.

Choice A is not correct because *twice* is a multiple number which should be followed by *as much as* with the noncount noun *land*. The phrase *twice as many kilometers* would also be correct because *kilometers* is a count noun that is used with *many*.

Practice 16

First, read the sentence. If it is correct, put a check mark beside it. ✔
If it is incorrect, edit the sentence.
Then check your answers with those in the Answer Key on page 370.

1. Within the wolf pack, the higher the rank, the taller the wolf stands, holding its head, ears, and tail erect to intimidate the others.

2. Very high mountains, like Pikes Peak, can be seen from as many as 80 miles away, which is about the distance to Limon, Colorado.

3. One colony of penguins can number as many 60,000 birds.

4. Almost of the traditional developmental psychologists have focused on childhood.

5. Although caffeine is a stimulant, the more coffee the subjects drank in the study, the less energy they reported after one hour.

★★★**Bonus:** iBT Speaking, Writing, ITP Essay
These are structures that you can use on tasks that require comparisons on the Speaking section and Writing sections.

Structure Strategy 17

Recognize noun endings in word families

Although it is usually very easy to identify parts of speech, word families can be confusing. Word families are groups of words with similar meaning and spellings. Each word in the family is a different part of speech. For example, *agreement* is a noun; *agreeable* is an adjective; *to agree* is a verb. The endings of words can help you classify the part of speech.

Nouns Derived from Verbs

Verb	Ending	Noun
store	-age	storage
accept	-ance	acceptance
insist	-ence	insistence
agree	-ment	agreement
authorize	-sion/-tion	authorization
research	-er	researcher

Nouns Derived from Adjectives

Adjective	Ending	Noun
convenient	-ce	convenience
redundant	-cy	redundancy
opposite	-tion	opposition
soft	-ness	softness
durable	-ity	durability

EXAMPLES

Unless protected areas are maintained, the Bengal tiger, the blue whale, and the California condor face _______ of extinction.

Ⓐ possible
● the possibility
Ⓒ to be possible
Ⓓ possibly

Only Choice B provides a noun object after the verb *face*. None of the other choices provide a noun as the part of speech. Choice A is an adjective. Choice C is an infinitive. Choice D is an adverb.

<u>The develop</u> <u>of hybrids</u> <u>has increased</u> <u>yields</u>.
A B C D

Choice A is not correct because *develop* is a verb, not a noun that can be used as a subject preceded by the article *a*. *The development* is the correct noun subject of the sentence.

Practice 17

First, read the sentence. If it is correct, put a check mark beside it. ✔
If it is incorrect, edit the sentence.
Then check your answers with those in the Answer Key on page 370.

1. Although it is most often associated with the mining of diamonds, South Africa is the world's largest produce of platinum.

2. A merger is an agree for two firms to continue in business as a single new company, rather than to remain separately owned and operated.

3. Because frogs drink and breathe through their skin, they are vulnerable to air and water pollution.

4. In 1884, a system of 24 standard time zones was adopted to calculate the rotate of the Earth by 15 degrees each hour.

5. The authorization for most computer security systems is based on a two-step process to verify the user's identity.

★★★**Bonus:** iBT Speaking, Writing, ITP Essay
Using the correct parts of speech improves scores on both the Speaking section and the Writing sections.

Structure Strategy 18

Identify endings for adjectives and adverbs in word families

Adjectives are sometimes derived from nouns and adverbs from adjectives. Adjectives modify nouns and adverbs modify verbs. Remember that each word in a word family is a different part of speech. For example, *possibility* is a noun; *possible* is an adjective; *possibly* is an adverb. The endings of words can help you classify the part of speech.

Adjectives Derived from Nouns

Noun	Ending	Adjective
possibility	-able/-ible	possible
intention	-al	intentional
distance	-ant	distant
frequency	-ent	frequent
juice	-y	juicy
interest	-ing	interesting
heat	internal change	hot

Adverbs Derived from Adjectives

Adjective	Ending	Adverb
efficient	-ly	efficiently
full	-ly	fully
perfect	-ly	perfectly
quick	-ly	quickly
strict	-ly	strictly

EXAMPLES

A beautiful white horse with one spiral horn, the unicorn is featured _______ in medieval art.

Ⓐ wide

Ⓑ widest

● widely

Ⓓ wideness

Choice C is an adverb that modifies the verb *featured* in the sentence. Choices A and B are adjectives and Choice D is a noun, none of which may be used to modify a verb.

No one knows <u>exactly</u> (A), but evidence <u>strong</u> (B) suggests that the Indus Valley civilization in Asia was using a <u>written</u> (C) language as <u>early</u> (D) as 3,500 B.C.E.

Choice B is not correct because the verb *suggests* requires an adverb modifier. *Strong* is an adjective. *Strongly* is an adverb.

Practice 18

First, read the sentence. If it is correct, put a check mark beside it. ✔
If it is incorrect, edit the sentence.
Then check your answers with those in the Answer Key on page 370.

1. The hot generated inside the Earth was much greater during the early history of the planet, due in large part to impacts, gravitational pull, and radioactive decay.

2. Diet, stress, and heredity are a few of the triggers for frequent migraine headaches.

3. It is currently estimated that our solar system orbits the galactic center once every 250 million years.

4. According to social learning theory, exposure to aggressive role models on television and in films can great influence violent behavior.

5. The use of mixed media can be more interest than that of other more traditional styles of artistic expression.

★★★★**Bonus:** iBT Reading, Speaking, Writing, ITP Essay
Studying word families will help you to extend your vocabulary for all of the sections of the TOEFL.

Structure Strategy 19

Make sure that the correct subject follows a verbal modifier

Introductory verbal modifiers introduce and modify the subject and verb in the main clause of a sentence. They can be *-ing* forms, *-ed* forms, or infinitives. They are usually separated from the main clause by a comma. The verbal modifier should immediately precede the noun that it modifies. Incorrect answers often present passives or nouns that are not logically modified by the introductory verbal modifier in the sentence.

EXAMPLES

_______ , air traffic controllers guide planes through conditions of near zero visibility.

Ⓐ They talk with pilots and watch their approach on radar

● Talking with pilots and watching their approach on radar

Ⓒ Talk with pilots and watch their approach on radar

Ⓓ To talk with pilots and watch their approach on radar

Only Choice B provides an introductory verbal modifier that logically refers to the subject *air traffic controllers*.

<u>Have designed</u> his <u>own</u> plane, *The Spirit of St. Louis*, Lindbergh <u>flew</u> from Roosevelt
A B C

Field in New York across the ocean to Le Bourget Field <u>outside Paris</u>.
D

Choice A is not correct because the introductory verbal modifier is not an *-ing* or an *-ed* form. *Having designed his own plane* would be correct. The subject, *Lindbergh*, is logically placed after the introductory verbal modifier.

Practice 19

First, read the sentence. If it is correct, put a check mark beside it. ✔
If it is incorrect, edit the sentence.
Then check your answers with those in the Answer Key on page 371.

1. Known as the Painted Lady in some parts of the world, it is between Africa and Europe where the brightly colored Red Admiral butterfly migrates.

2. To be finding naturally in many kinds of food, minerals are essential to maintain healthy bones, a strong immune system, and normal blood pressure.

3. Founded in 1636, Harvard was originally called the New College, but it was not until 1639 when John Harvard bequeathed his library and half of his estate to the school that the college began to provide higher educational programs.

4. Embraced primarily as a reaction against the Age of Reason, artists introduced Romanticism not as a style of art so much as a pursuit of beauty intended to arouse an emotional response.

5. Running at speeds of up to thirty miles per hour, the Arctic hare can keep up with a car in city traffic.

★★**Bonus:** iBT Writing, ITP Essay
Introductory verbal modifiers can be used on the Writing sections to provide variety in sentence structure.

Structure Strategy 20

Look for similar structures in lists and after correlative conjunctions

Parallel structure is the use of the same grammatical structures to express ideas of equal importance. When you see a list or a correlative conjunction, look for parallel structures.

Correlative Conjunctions

both . . . and . . . as well as . . .
not only . . . but also . . .
either . . . or . . .
neither . . . nor . . .

EXAMPLES

Incorrect:
Schizophrenia, a behavioral disorder typified by a fundamental break with reality, may be triggered by genetic predisposition, stressful, drugs, or infections.
Correct:
Schizophrenia, a behavioral disorder typified by a fundamental break with reality, may be triggered by genetic predisposition, stress, drugs, or infections.

Incorrect:
When the atmosphere of early Earth had more carbon dioxide, microorganisms probably used either hydrogen or it was hydrogen sulfide as a source of electrons instead of water.
Correct:
When the atmosphere of early Earth had more carbon dioxide, microorganisms probably used either hydrogen or hydrogen sulfide as a source of electrons instead of water.

Incorrect:
Homer wrote not only the *Iliad* but also he wrote the *Odyssey*, epic poems that had a major influence on the history of literature.
Correct:
Homer wrote not only the *Iliad* but also the *Odyssey*, epic poems that had a major influence on the history of literature.

Incorrect:
Children are neither viewed as outsiders nor they are seen as emergent members of the social structure because they participate almost immediately after birth in society.
Correct:
Children are neither viewed as outsiders nor seen as emergent members of the social structure because they participate almost immediately after birth in society.

Incorrect:
One of the most expensive spices by weight, saffron grows best in strong sunlight, well-drained beds, and organically fertilized.
Correct:
One of the most expensive spices by weight, saffron grows best in strong sunlight, well-drained beds, and organically fertilized soil.

Practice 20

First, read the sentence. If it is correct, put a check mark beside it. ✔
If it is incorrect, edit the sentence.
Then check your answers with those in the Answer Key on page 371.

1. The Bamboo Lemur, a critically endangered species, feeds almost exclusively on bamboo, eating not only the shoots but the pith and leaves.

2. The grieving process may be triggered by a number of situations, including a loved one dies, a chronic condition that affects the quality of life, or even the loss of a job.

3. Gold is a highly efficient conductor of electricity, which makes it useful in small amounts in cell phones, global positioning systems, and calculators.

4. Neither Benjamin Franklin nor Thomas Jefferson was in attendance at the First Continental Congress.

5. At more than 21,000 meters, the highest known mountain in the solar system is not on Earth but it is Mars.

★★★**Bonus:** iBT Speaking, Writing, ITP Essay
Your use of parallel structure on the Speaking section and the Writing section of the iBT and the ITP essay will impress raters.

6
STRATEGIES FOR THE WRITING SECTION

OVERVIEW OF THE iBT WRITING SECTION

The Writing section tests your ability to write essays similar to those that you would write in college courses. During the test, you will write two essays. You will be required to type them.

The Integrated Essay. First you will read an academic passage and then you will listen to a lecture on the same topic. You may take notes as you read and listen, but notes are not graded. You may use your notes to write the essay. The reading passage will disappear while you are listening to the lecture, but the passage will return to the screen for reference when you begin to write your essay. You will have 20 minutes to plan, write, and revise your response. Typically, a good essay for the integrated topic will require that you write 150–225 words.

The Independent Essay. You will read a question on the screen. It usually asks for your opinion about a familiar topic. You will have 30 minutes to plan, write, and revise your response. Typically, a good essay for the independent topic will require that you write 300–350 words.

A clock on the screen will show you how much time you have left to complete each essay.

OVERVIEW OF THE ITP WRITING SECTION

The Writing section of the ITP is a separate test, the Test of Written English (TWE). It is administered at the end of the ITP and is usually required as part of the test. The TWE tests your ability to write an essay about a general topic. During the test, you will write one essay. You can choose to type or handwrite your essay.

You will have 30 minutes to plan, write, and revise your response. Typically, a good essay for the TWE topic will require that you write 300–350 words.

TIPS FROM SUCCESSFUL TOEFL STUDENTS

WHAT TO DO ON THE WRITING SECTION

- Be sure that you understand the writing question before you start. Read it carefully so that you don't answer off topic.
- When you are taking practice tests to prepare for the Writing section, disable the spell check on your computer. The spell check feature is not available on the TOEFL keyboard.
- If you are taking an English class, keep going to class. Ask your teacher to evaluate your practice essays. Most of your English teachers will be glad to help you, even if they are not teaching a writing course.
- If you ask friends to help you or you hire a tutor, be sure that they understand the scoring guides for the TOEFL. You need feedback on the specific points that raters will check.
- Don't worry when you see other people finishing the test. They aren't getting points for leaving first!

Writing Strategy 1

Improve your typing skills

You must type your essays on the iBT. If you improve your typing skills, you can focus on the content of your essay and type faster. Here are some tips:

1. *Type with both hands.* If you use only one or two fingers to look for the keys, you could waste valuable time.
2. *Become familiar with a QWERTY keyboard.* TOEFL Centers use a standard English keyboard, named for the keys in the top left row under the numbers.
3. *Practice typing on several desktop computers.* It will be helpful to have experience typing on different computers because you will be more comfortable when you use an unfamiliar desktop in the TOEFL Center.

EXAMPLE KEYBOARD

Practice 1

First arrange to use a desktop computer with a QWERTY keyboard. Then, using both hands, copy the following essay. Time yourself. Type as quickly as possible without making mistakes because of speed. Then check your essay for errors. You should set a goal: 15 minutes with 5 errors. Compare your typing score with the scores on the chart in the Answer Key on page 372.

Essay

In some countries, students are encouraged to work part time while they are attending college. Although a case could be made for students to use all of their free time studying, in my view, working part time is a good idea because it allows young people to gain experience in the workplace, makes a contribution to their living expenses, and provides an incentive for continuing their educations.

Even a menial job like serving food or cleaning an office allows students to establish a routine and practice basic skills. For example, showing up on time, completing a task, and interacting with an employer and coworkers are all good skills for any job. Students without work experience will not have the advantage of having learned and established habits that can be easily transferred to other jobs. Even a part-time job can be entered on a resume to prove that an applicant understands the work environment.

In addition, the income from a job can be important to both the student and the family. Even when a student has earned a scholarship or receives financial aid, living expenses can be very burdensome. Room, board, incidentals, and books often cost more than the budget that the family has planned in advance. A part-time job can make the difference between asking for more money from home or cutting back on basic necessities like nutritious food or a safer place to live. Working indicates a willingness to assume some of the responsibility for the costs associated with a college education.

In my experience, the kind of employment available for those who do not yet have their degrees is not the job that most students would aspire to achieve as a permanent position. Part-time jobs for students usually include low-level tasks that would not be attractive to someone with a college degree. Working in an entry-level job can provide motivation for students to finish their degrees and qualify for better employment opportunities. Most college students will learn very quickly that they don't want to work in their first jobs forever.

In conclusion, I support part-time employment for college students to provide them with work experience, extra money, and incentives to complete their degrees.

Writing Strategy 2

Anticipate familiar topics

You will be asked to write two essays on the iBT Writing section, including an integrated essay about an academic topic and an independent essay about a familiar topic. The ITP essay question also requires an opinion about a familiar topic. Familiar topics for the iBT independent essay and the ITP essay include four categories:

1. Description and Explanation
2. Agreement—Disagreement
3. Preference
4. Condition

Review the topics in the examples below. You may not see exactly the same topics on your test, but the review will help you begin organizing your thoughts and opinions before the test, and you will probably not have to spend as much time thinking about your ideas when you start writing your essay.

Example Topics

Description and Explanation

Describe an ideal friend.

Describe a good roommate.

Describe a good neighbor.

Describe a perfect husband or wife.

Describe your ideal job.

Describe a good parent.

Describe a beautiful city.

Describe a perfect vacation.

Describe an excellent teacher.

Describe your favorite book.

Describe your favorite holiday.

Describe the most important invention.

Describe your favorite gift.

Describe an important symbol of your country.

Describe the most important quality for success.

Agreement/Disagreement

Life is easier now than in the past.

Computers are a positive influence.

Class attendance should be required.

Difficult experiences can provide important lessons.

Foreign languages should be taught to young children.

Boys and girls should attend separate schools.

It is always important to be on time.

A first impression is usually correct.

School uniforms should be required in secondary school.

Playing sports teaches us about life.

Pets should be treated like family members.

Celebrities deserve the high salaries that they receive.

Experience is the best teacher.

It is a good idea to participate in activities at school.

Music is important in the lives of most people.

Preference

Some students prefer a traditional classroom. Others like to take classes online.
Which type of class do you prefer?

Some people spend most of their money when they receive it. Other people save as much as possible.
What do you think is better?

Some teachers lecture for most of the class. Other teachers spend part of the class in small group discussions.
Which method do you think is better?

Some people use physical activity to alleviate stress. Other people prefer to escape in a good book or watch a movie.
Which plan do you use in stressful situations?

Some people like to work outdoors. Other people like to work indoors.
Which kind of work do you prefer?

Some students prefer to lead a group and others prefer to be part of a group without a leadership position.
What do you prefer to do?

Some parents enroll their children in activities and special classes like music and tennis. Other parents believe that children should spend their free time playing.
What do you think is better for children?

Some people like to read fiction. Other people prefer to read non-fiction.
Which type of books and materials do you prefer to read?

Some employers hire their employees on the basis of their work experience. Other employers think that a degree is more important.
Which kind of employee would you prefer to hire?

Some people support building a factory in your town. Others oppose it.
What do you think about the plan?

Some students prefer to live in a dormitory on campus. Other students prefer to find an apartment close to school.
Which living situation would you prefer?

Some people like to communicate by e-mail and text message. Other people prefer to talk by phone or face to face.
Which kind of communication do you prefer?

Some students like to work on projects independently. Other students enjoy being part of a group effort.
Which option for a project would you prefer?

Some students would like to be accepted to a large university. Other students would rather go to a small college.
Where would you prefer to study?

Some people enjoy traveling with friends. Other people prefer traveling alone.
How do you like to travel?

Condition

If you could visit any country in the world, where would you go?

If you could change one thing about yourself, what would it be?

If you could change one thing about your home town, what would you do?

If you could meet anyone who is alive today, which person would you want to meet?

If you could go back in time to any place and year, which period would you choose?

If you could show tourists any place in your country, where would you take them?

If you received a large sum of money, how would you spend it?

If you could choose to take a course for fun, which course would you take?

If you could give a prize to any person, who would receive it?

If you could select one object to represent your country in a world's fair, what would it be?

If you could give one piece of advice to a younger student, what would it be?

If you could make one important change in the world, what would it be?

If you needed to borrow money, would you use a bank or ask a friend?

If you could make a large contribution to a charity, which one would you choose?

If you could start a business, what kind of business would it be?

If you could predict the future, what do you think life will be like fifty years from now?

Practice 2

First read the topics from the examples. Then try to form an opinion for each topic. It is not necessary to write anything down, but you may take notes if it helps you to think. This is called *brainstorming* in English. It means that you are getting ideas together in your mind. Don't rush through the list. Consider your opinions. Next, complete the following sentences as quickly as you can. If you have already thought about your opinions, you should be able to respond without further brainstorming. Finally, compare your answers with the example answers in the Answer Key on page 372.

1. My ideal job would be ________________________________.
2. I think that an excellent teacher ____________________________________.
3. My favorite book is ____.
4. The most important invention in the world so far is ____________________________.
5. The Internet is a __________________________ influence.
6. In my opinion, class attendance __________________ required.
7. In my view, __________________________ is a good way to alleviate stress.
8. I prefer a __________________________ classroom.
9. If I could meet any person, I would like to meet ______________________________.
10. If I could change one thing about my home town, I would __________________________

    ____.

Writing Strategy 3

Choose the opinion that is easiest to support

When you read the independent essay question, choose an opinion that you can support with three reasons or examples that come immediately to mind. Do not waste time trying to decide which opinion will be better. You are not obligated to tell the truth about your point of view. The raters will not grade your opinion. They will be evaluating your ability to *support* your opinion. That means that you do not have to choose an opinion that you agree with or one that you think the raters might agree with.

EXAMPLES OF OPINION ESSAYS

QUESTION

Agree or disagree with the following statement: Experience is more important than education when hiring a new employee.

Disagree: Excellent Score

Although it can be argued that experience is important to success in the workplace, I disagree that it is more important than education when hiring a new employee. Three reasons persuade me to put forward this opinion. First, new ideas are very important to any company; however, if only experienced employees are hired, the potential for new ideas could be more limited. Young, inexperienced employees are more likely to have exposure to recent information in their educational programs. Moreover, they may be more flexible about trying something new because they are less committed to maintaining the status quo. New ideas and new ways of responding to a changing environment are essential to the continuing success of a company in a competitive market.

In the second place, training is usually a major component of a successful company. Experienced workers often resent training because they want to continue using their prior knowledge as the basis for their future contributions to the workplace. In contrast, less experienced employees are more open to training and will probably implement some of the ideas that they learn in the training sessions. Furthermore, the education that they bring to the training will be valuable. New employees with a higher level of education may be better able to learn than employees who are experienced but have not been in a learning environment for a long time.

Finally, workers with less experience often command lower salaries. In some cases, it is actually possible to hire two young employees instead of one more experienced employee for the same expenditure. If a company is considering labor costs, then well-educated but less experienced employees may be a better short-term investment. It can also be argued that two employees would be a better long-term investment, since experience can be acquired on the job.

For a company with a view to implementing new ideas, providing training to support their employees in a changing marketplace, and increasing their workforce, a well-educated employee with less experience may prove a better choice, especially if the candidate comes with enthusiasm, a desire to cooperate with other employees, and a good work ethic, all of which will contribute to success regardless of education or experience.

Agree: Excellent Score

In my view, experience is a more important consideration than education when hiring a new employee. Three reasons persuade me to put forward this opinion. In the first place, an experienced worker has an employment history that can be verified. Experience implies that the candidate has worked for one or more companies and will be able to include useful references in the application package. In the case of an inexperienced employee, only personal or school-related references will be available. Success in school is not always a predictor of success in the workplace. For example, Einstein failed math in school. A company that wants to be more certain of the potential contribution that an employee can make on the job should consider the advantages of information about previous work experience.

Second, inexperienced employees can waste time on the job because they are not trained and have not yet confronted common problems. When experienced employees are trying to find solutions, they can refer to similar situations and eliminate options that have failed in the past. Even well-educated employees would probably have to use a trial-and-error approach to determine how to handle a problem. Moreover, since time is money, the quicker response by experienced employees would be reflected on the bottom line.

Finally, experienced employees tend to be older and may be able to serve the company in the role of trainer or mentor in addition to the position for which they are hired. Even in a less-official capacity, older workers can provide good examples to younger workers. By observation, a less-experienced coworker can learn many useful professional and personal skills and shortcuts from a more experienced employee.

For a company that needs to be sure that a candidate will be able to perform, think through a problem quickly and choose a solution that has worked in the past, and serve as a coach within the organization, an experienced employee would prove a better choice, especially if the candidate comes with enthusiasm, a desire to cooperate with other employees, and a good work ethic, all of which will contribute to success.

Practice 3

First, read the statement in the question. Next, in one minute, list as many reasons as you can to agree with the statement. Then, in one minute, list as many reasons as you can to disagree with the statement. Which opinion would you choose for each question? Compare your answers with the example answers in the Answer Key on page 372.

QUESTIONS

1. Childhood is the best time of life.

Agree	Disagree
______________	______________
______________	______________
______________	______________
______________	______________

2. Science is the most important subject in school.

Agree	Disagree
______________	______________
______________	______________
______________	______________
______________	______________

3. Computers have improved communication.

Agree	Disagree
______________	______________
______________	______________
______________	______________
______________	______________

4. Playing games is a waste of time.

Agree	Disagree
____________________	____________________
____________________	____________________
____________________	____________________
____________________	____________________

5. Endangered species should be protected.

Agree	Disagree
____________________	____________________
____________________	____________________
____________________	____________________
____________________	____________________

6. Parents are the best advisors for their children.

Agree	Disagree
____________________	____________________
____________________	____________________
____________________	____________________
____________________	____________________

7. It is better to read non-fiction books than novels.

Agree	Disagree
____________________	____________________
____________________	____________________
____________________	____________________
____________________	____________________

8. First impressions of people are usually correct.

Agree	Disagree
________________	________________
________________	________________
________________	________________
________________	________________

9. Celebrities receive too much money for their work.

Agree	Disagree
________________	________________
________________	________________
________________	________________
________________	________________

10. Exercise is the best way to manage stressful situations.

Agree	Disagree
________________	________________
________________	________________
________________	________________
________________	________________

★**Bonus:** iBT Speaking

Tasks 1 and 2 in the iBT Speaking require that you choose an opinion to support.

Writing Strategy 4

Include an outline sentence

When you write your introduction, you should include not only a thesis statement but also an additional sentence that briefly outlines your essay. Include three main ideas in the outline sentence. For maximum impact, you should express these ideas using the same grammatical structures. Put the outline sentence at the end of the first paragraph.

Examples

Thesis Statement

Although most people must work in order to support themselves and their families, work provides several other benefits for employees.

Outline Sentence

In my view, the most important benefits are a regular schedule, a network of colleagues and friends, and a sense of purpose.

Thesis Statement

Hobbies improve our lives in a number of ways.

Outline Sentence

To mention only three, hobbies are a good form of relaxation, a way to meet people with a common interest, and a potential source of additional income.

Thesis Statement

The most interesting place that I have visited on vacation is London.

Outline Sentence

I enjoyed it because I had the opportunity to visit the most famous museums, tour several historic landmarks, and see a play at the Globe Theater.

Thesis Statement
Although attending school in a foreign country is a wonderful opportunity, it also presents new students with problems.

Outline Sentence
Perhaps the most common problems are establishing a new living situation, adjusting to different expectations in school, and overcoming homesickness.

Thesis Statement
Maintaining good health is important at any age.

Outline Sentence
As a relatively young person, I try to stay healthy by eating a balanced diet, including exercise in my daily routine, and getting enough sleep.

Practice 4

First, read the thesis statement for each essay. Then underline the beginning of the sentence that outlines the essay. Finally, include three main points to complete the outline sentence. Be sure to use parallel structure to list the main points. Compare your answers with the example answers in the Answer Key on page 374.

1. My older brother is the person that I most admire. I feel that way about him because

   ____

   ____

   ____

2. If I were to receive a million dollars, I would be amazed. My plan for using the money would include

   ____

   ____

   ____

3. The announcement that a factory will be built on the former site of a park in my hometown is not good news. I am distressed to hear about this development because

   ____

   ____

   ____

4. I would rather live in the city than in the country because of the opportunities available in an urban setting. Perhaps the most important advantages are

   ____

   ____

   ____

5. Although a case could be made for small families, I would rather have a large family. A large family is important because

   ____

   ____

   ____

6. When I choose a friend, I look for a number of characteristics, but probably the most important to me is trustworthiness. I value this quality in a friend because

___

___

___

7. While I could discuss several ways that the Internet has changed the world, three ways seem most significant. The Internet has

___

___

___

8. The government has passed legislation to ban smoking in public places. I think that this law is appropriate because

___

___

___

9. Although school is most often associated with education, there are other ways to learn valuable lessons. Three ways that seem worth exploring are

___

___

___

10. Given the opportunity to work for a company or start my own business, I would definitely choose to be a small business owner. The idea of starting my own business is appealing to me because

___

___

___

★**Bonus:** iBT Speaking
Although the introductions to the Speaking tasks must be very short, a brief outline sentence is still useful as a first or second sentence because an outline sentence is a direct statement.

Writing Strategy 5

Use the number 3 to organize your essay

You already have three reasons to support your opinion in the outline sentence. Now write one sentence to expand on each of the three reasons in your outline sentence. Use each of these three sentences as the first sentence in the three paragraphs in the body of your essay. It is possible to write a good essay that does not conform to the usual number 3, but some raters will look for this traditional organization.

EXAMPLE NOTES

Description and Explanation

Describe an ideal friend.

For someone to be my ideal friend, she has to have compatible interests and personality traits. I would prefer a friend who is ambitious, has a strong work ethic, and is able to adapt.

1. My ideal friend must have a desire to improve herself.
2. She should also demonstrate a willingness to work hard.
3. Her ability to learn from mistakes would also be an important trait.

Agreement/Disagreement

Life is easier now than in the past.

I agree that life is easier now than it was in the past. This is the case primarily because of new machines, better medicine, and increased opportunities.

1. Machines have taken the place of manual labor.
2. Modern medicine has improved general health conditions.
3. Educational and employment opportunities have expanded.

Preference

Some students prefer a traditional classroom. Others like to take classes online.
Which type of class do you prefer?

I prefer to take classes online instead of attending a traditional class. Some of the advantages include better scheduling, greater communication, and lower expenses.

1. The class schedule can be adjusted to fit my routine.
2. I can communicate more effectively with the teacher and other students.
3. The cost of books and lab materials is lower for online classes.

Condition

If you could change one thing about yourself, what would it be?

If I could change one thing about myself, I would instantly acquire more patience than I currently possess. This change would make a positive difference in my interactions with people, my physical and mental condition, and my choices.

1. Patience would improve my relationships.
2. Being more patient would help to restore my health.
3. If I were more patient, I would make better decisions.

Practice 5

Choose one of the topic sentences below and complete an outline statement. Then write three reasons that support your opinion. Remember, there are no right or wrong opinions, but you need to be able to support the opinion you have. The three reasons that you write for the topic can be used as the first sentence for each of the three paragraphs in the body of your essay. Compare your answers with the example answers in the Answer Key on page 375.

QUESTION

Some people like to study alone. Other people prefer to study in a group. Which do you prefer? Use specific reasons to support your opinion.

Although many people prefer to study in a group, I would rather study alone. Studying by myself is more effective because

1. ____

____

____

2. ____

____

____

3. ____

____

____

Although many people prefer to study alone, I would rather study in a group. Studying in a group is more effective because

1. ______________________________

____

____

2. ______________________________

____

____

3. ______________________________

____

____

★**Bonus:** iBT Speaking

Tasks 1 and 2 in the Speaking section of the iBT require that you support your opinion. The number 3 is a traditional strategy for speaking as well as for writing.

Writing Strategy 6

Include facts, examples, and experience for supporting details

You need supporting details to explain and expand on the major points in your essays. Some supporting details are factual evidence, but all supporting details do not have to be facts. Some details can be examples or personal experience.

Example of Supporting Details

Attendance should not be part of the grading system in university courses. In my opinion, the policy is misguided because it does not encourage teachers to prepare interesting classes, it does not take into account the differences in students, and it does not set a positive example for students.

Major Point 1

The policy serves the interests of the teacher, not the students.

Supporting Detail Experience

Although many professors plan interesting classes, some professors do not provide additional information. In my experience, the lectures in about half of my undergraduate classes were very repetitive. The information that the professor provided was also available in the textbook or in other course materials online. Since no value was added to the course by attending classes, the policy seemed to be more punitive than positive. It would have been possible to earn an A without attending classes if the attendance policy had not been instituted. Therefore, the policy seemed to support the teachers who did not bother to prepare their classes well. It did not seem to be in the best interest of students.

Major Point 2

By making attendance mandatory, the professor is assuming that all students are alike.

Supporting Detail Evidence

However, research on learning styles confirms that some students need the structure provided by a classroom environment, but others learn better by reading and researching on their own. For students who are independent learners, the classroom time is not well spent, but they do not have the option to learn in a different way without jeopardizing their grades. Instead of spending their time productively, students who learn better by using resources in labs and libraries have to interrupt their study time to attend class.

Major Point 3

Finally, mandatory attendance sends the wrong messages to students.

Supporting Detail Example

For example, when students who attend class are rewarded for sitting in their seats and obeying the rules, not for making an effort to learn in the best way that they can, then the lesson is more about conformity than the content of the class. Students get the message that sitting obediently is more important than excelling on an exam.

Practice 6

First, read the three major points for the essay. Then add one supporting detail under each point. Use evidence, examples, or experience. Compare your answers with the example answers in the Answer Key on page 375.

Attendance should be part of the grading system in university courses. In my opinion, the policy is appropriate because many students will not make mature choices, they profit from a classroom review of the material, and they prepare for other life experiences by respecting an attendance requirement.

Major Point 1
Because many university students are still immature, the requirement encourages them to stay on track.

Supporting Detail

____

____

____

Major Point 2
Even if some of the lectures are repetitive, it is good to review material that students have read in the book and online.

Supporting Detail

____

____

____

Major Point 3
Since mandatory attendance in classes is not very different from obligations in the workplace, the policy prepares students for life after they leave the university.

Supporting Detail

____

____

____

Writing Strategy 7

Choose appropriate transition words for written English

You should use informal transition words when you respond to the questions in the Speaking section of the TOEFL, but you need to use formal transition words when you write an essay. The most common transition words are listed below. Use a semicolon before each formal transition word and a comma after it.

Informal	Formal
but	; however,
and	; furthermore,
and	; moreover,
still	; nevertheless,
so	; therefore,

Examples

Spoken English: Many insects die during the winter months, but a small number of them survive to repopulate the nest.

Written English: Many insects die during the winter months; however, a small number of them survive to repopulate the nest.

Spoken English: In some traditional cultures, rites of passage mark an individual's transition from child to adult status, and it signals the time for separation from the mother.

Written English: In some traditional cultures, rites of passage mark an individual's transition from child to adult status; furthermore, it signals the time for separation from the mother.

Spoken English: Minoan architecture included a system of plumbing, and they engineered light wells to illuminate underground rooms.

Written English: Minoan architecture included a system of plumbing; moreover, they engineered light wells to illuminate underground rooms.

Spoken English: Barrier islands are smoothed by waves on their seaward sides. Still their landward margins are irregular because storm waves carry sediment over the island and deposit it in lagoons.

Written English: Barrier islands are smoothed by waves on their seaward sides; nevertheless, their landward margins are irregular because storm waves carry sediment over the island and deposit it in lagoons.

Spoken English: A mezzo soprano has a range lower than a typical soprano but somewhat higher than an alto, so a mezzo soprano is often cast in the role of an older woman or a supporting character.

Written English: A mezzo soprano has a range lower than a typical soprano but somewhat higher than an alto; therefore, a mezzo soprano is often cast in the role of an older woman or a supporting character.

Practice 7

First, read the script for the spoken response. Then write the response using the more formal transition words that would be appropriate for an essay. Compare your answers with those in the Answer Key on page 376.

Spoken English: In selective listening experiments, subjects are presented simultaneously with two spoken messages, but most of the time they are able to recall the message that they are instructed to remember and disregard the other one.

Written English: ____________________

____

____

____

Spoken English: When the Romans invaded Britain, they had to overcome the problem of staging an attack across the English Channel, and they were faced with dissatisfaction within the ranks of their armies.

Written English: ____________________

____

____

____

Spoken English: The Sophists believed that people should take care of themselves, so they advocated a system of education that promoted individual accountability.

Written English: ____________________

____

____

____

Spoken English: Earth's magnetic field is generated within its molten core. Still several questions remain about the changes in the core's convection.

Written English: ____

____

____

____

Spoken English: Japan, Taiwan, and Mexico all have early warning systems to provide a short time for the population tc take shelter, and many countries are now developing cell phone programs with a warning ring.

Written English: ____

____

____

____

★**Bonus:** iBT Speaking

Learning informal transition words as well as formal transition words will prepare you for the Speaking section. Use informal transition words when you respond to the Speaking tasks.

Writing Strategy 8

Restate the major points to summarize the reading passage

You will have three minutes to read the 250- to 300-word passage that appears before the lecture for the integrated essay. Summarize the reading passage in the first few sentences of your essay. Remember that your summary should not be as long as the original. Include only the major points and the most important details. You will usually find the major points in the first sentences of each paragraph. The content must be accurate, but you should be able to restate the major points of the reading passage in about 100 words.

EXAMPLE OF READING PASSAGE

From geological evidence, it is acknowledged that at least four massive glaciers advanced and receded over the North American hemisphere during the Pleistocene period. During the last Ice Age around 10,000 years ago, at the end of the Pleistocene period, a large portion of North America was covered by ice as thick as 6,500 feet. Temperatures on land were about 25 degrees colder than they are now. About 18,000 years ago, Laurentide, the last large glacier, covered most of Canada and extended as far south as Pennsylvania in the Midwestern United States, an area of more than 5,000,000 square miles.

As the Laurentide glacier moved south along what is now the Great Lakes, the ice sheets sheared off mountaintops and carved deep valleys in a process called glacial erosion. Their leading edges left high ridges behind. The advance of the glacier enlarged the system of river valleys that already existed and scoured out enormous basins.

Thousands of years later, as the climate began to warm, the ice sheets began to retreat and melt, and the basins carved by the glacier were filled with water. In this way, the Great Lakes were formed, along with the St. Lawrence River Valley. The original lakes were much larger than they are today, but their basic shapes were already evident and the drainage patterns were established for the five individual lakes that are currently distinguished. They are connected to each other through channels, forming one system of water that flows from Lake Superior through the entire system, comprising the largest group of fresh water lakes on Earth.

EXAMPLE OF SUMMARY

During the Pleistocene period, four huge glaciers flowed and retreated in North America. Moving south over the area now known as the Great Lakes, the Laurentide glacier changed the topography of the region, leveling mountains and gouging out deep basins. When the climate warmed several thousand years later, the ice started to melt, filling the basins with water. Thus, the Great Lakes and the St. Lawrence Seaway were formed.

Practice 8

First, read the passage, paying special attention to the first sentences in each paragraph. Then write a summary about 100 words in length. Compare your answer with the example answer in the Answer Key on page 377.

An herb is a plant that is valued for medicinal or other beneficial qualities because it contains chemicals that have proven effective in the treatment of various illnesses. Many herbal treatments were developed by observing the effects of plants on animals and by experimentation. Knowledge of herbal cures was passed down from one generation to the next in oral histories.

Although herbal remedies do not enjoy the same level of recognition among the scientific community that pharmaceuticals do, herbal medicine has a long tradition in almost every culture. As early as 2,735 B.C.E., a Chinese emperor wrote an extensive volume on herbs, which is still referred to today. The ancient cultures of Egypt and India also published texts that listed medicinal herbs and their uses, and the records of King Hammurabi of Babylon included prescriptions for medicinal herbs. Native people in both North and South America used plant life for both ceremonial and natural cures. Furthermore, Europeans relied on plants as medicine for many centuries. In the mid-seventeenth century, Nicholas Culpeper published *The English Physician*, which was one of the first home references for family medicine. It contained a listing of herbal remedies and is still useful today.

The World Health Organization (WHO) estimates that approximately 80 percent of the world's population uses herbal medicine as a primary form of treatment. WHO also released a report in which 120 medicines derived from plants were documented. Almost 75 percent of the plant derivatives were found to correlate very closely to modern treatments and native remedies. For example, salicylic acid, a substance that was used prior to aspirin, was originally derived from white willow bark and meadowsweet plants. Opium poppies were used for pain before morphine and codeine were developed from the plants, and cinchona bark was the source for quinine, a treatment that is still used for malaria. Plant compounds are still the main component of pharmaceutical medications prescribed for heart disease, high blood pressure, and chronic obstructive pulmonary conditions, including asthma.

Summary

★**Bonus:** iBT Speaking

Summarizing is an important strategy for both the integrated essay in the Writing section and the integrated tasks in the Speaking section of the iBT.

Writing Strategy 9

Substitute synonyms to paraphrase a sentence

If you copy directly from a reading passage or use words directly from a lecture, you can receive a 0 score on your essay. One easy way to paraphrase is to substitute synonyms for the vocabulary that you read or hear.

Examples

Although most scientists believe that animals have feelings, they also suspect that animal feelings are different from those of human beings.

Although most researchers think that animals exhibit feelings, they also assume that animal feelings differ from those of humans.

Whereas Latin served as the primary language of communication in Europe, Chinese was used in Asia and Arabic in the Islamic Empire.

Whereas Latin was the principal language for European intercultural communication, Chinese served that purpose in the Far East and Arabic in the Middle East.

In the late twentieth century, architects began to consider the impact of their buildings on the environment not only in terms of materials but also in terms of the energy that would be required to sustain them.

In the late 1900s, architects began to take into account the effect that their structures would have on the environment both in terms of resources and energy requirements to sustain them.

The jaguar has a short, compact body and a big head with teeth and jaws strong enough to drag prey two or three times its weight for great distances.

The jaguar is short and sturdy with a large head and powerful teeth and jaws that allow it to pull prey that weighs twice or even three times its weight for long distances.

Some of the prehistoric art found in caves include very early maps that show fields, paths, dwellings and other specifics, including urban settlements and even mountains, rivers, and volcanoes.

Some examples of prehistoric cave art consist of extremely ancient maps that identify areas under cultivation, paths, houses, and other details, including towns and topographical features.

Practice 9

First, read the original sentence. Then try to think of synonyms to paraphrase the underlined words or phrases and rewrite the sentence. Compare your answers with the example answers in the Answer Key on page 377.

1. It is remarkable that most individuals have the ability to remember seven discrete chunks of information.

2. No universal definition of music exists because the difference between music and noise is identified differently by various cultures.

3. Vascular plants first appeared during the Silurian period, and by the Devonian period, they had diversified and dispersed into many different land environments.

4. One of the oldest surviving examples of casting is a copper frog from 3,200 B.C.E., which was found in Mesopotamia.

5. Typically, only one-tenth of an iceberg is visible above the water, which makes it difficult to determine the shape of the part submerged.

6. Polaris, also called the North Star, appears to remain stationary in the northern hemisphere, while other stars and planets revolve around it.

7. In some states, an optional preschool year is offered to children of five to six years of age, while in other states, all five-year-olds attend a compulsory year of kindergarten.

8. Aboriginal Australians were believed to have first arrived on the Australian mainland by boat from the Indonesian Archipelago about 50,000 years before the Europeans.

9. Amphibious fish like the mudskipper can breathe out of water for <u>several</u> days and <u>move about</u> on land by <u>manipulating</u> their fins in a skipping motion.

____

____

____

____

10. Twins <u>experience</u> a <u>unique</u> language learning <u>situation</u> because, unlike singletons who learn from adults and older siblings, twins <u>receive</u> input from another child at the same <u>level</u> of language <u>acquisition</u>.

____

____

____

____

★**Bonus:** iBT Speaking
Paraphrasing is an important strategy for the integrated tasks on the iBT Speaking section as well as for the integrated essay on the iBT Writing section.

Writing Strategy 10

Report the content information

Use indirect statements and strong verbs to report the content information in the reading passage and the lecture. Choose verbs for your report based on the point of view expressed by the author of the reading passage and the speaker presenting the lecture—doubtful, neutral, certain, or negative.

Doubtful Verbs	Neutral Verbs	Certain Verbs	Negative Verbs
allege	describe	argue	contradict the theory/idea/concept
assume	explain	assert	challenge the theory/idea/concept
believe	illustrate	conclude	deny the theory/idea/concept
claim	indicate	confirm	dispute the theory/idea/concept
imply	mention	contend	reject the theory/idea/concept
propose	note	demonstrate	
suggest	observe	discover	
suppose	point out	find	
suspect	report	maintain	
think	say	prove	
	show		
	state		

EXAMPLES

Content

1. New volcanoes are forming all of the time, usually in the same areas where existing volcanoes have already been found.

Certain Report

The lecturer contends that volcanoes are forming all of the time, usually in the same areas where existing volcanoes have already been found.

Content

2. Evidence from Harappi documents an Indus Valley Civilization that had a written language as early as 3,500 B.C.E.

Neutral Report

The reading passage explains that evidence from Harappi documents an Indus Valley Civilization that had a written language as early as 3,500 B.C.E.

Content

3. The brain is the only organ that is more active at night than during the day.

Doubtful Report

The lecturer claims that the brain is the only organ that is more active at night than during the day.

Content

4. Cave art reduced images to the most essential, stylized forms and eliminated most of the details.

Neutral Report

The lecturer observed that cave art reduced images to the most essential, stylized forms and eliminated most of the details.

Content

5. Genetic drift may occur by chance fluctuations in the gene pool.

Doubtful Report

The reading suggests that genetic drift may occur by chance fluctuations in the gene pool.

Content

6. Social stratification is functional for society because it guarantees that the most able members will be assigned to the most demanding positions.

Negative Report

The reading passage rejects the theory that stratification is functional for society because it guarantees that the most able members will be assigned to the most demanding positions.

Content

7. Immigration from abroad was not significant until the 1820s.

Certain Report

The lecture confirms that immigration from abroad was not significant until the 1820s.

Content

8. A nebula, which is a cloud of dust and gas, can span across many light years.

Certain Report

The lecturer argues that a nebula, which is a cloud of dust and gas, can span across many light years.

Content

9. Polygraph testing is accurate about two-thirds of the time.

Neutral Report

The reading passage reports that polygraph testing is accurate about two-thirds of the time.

Content

10. The basic design of the drum has remained virtually unchanged for thousands of years.

Certain Report

The reading passage asserts that the basic design of the drum has remained virtually unchanged for thousands of years.

Practice 10

First read the sentence to understand the information. Next, write a sentence that reports the content. Then compare your answers with the example answers in the Answer Key on page 378.

1. Perceptions are based on larger units such as forms, figures, and contexts.
   Negative Report

   ____
   ____
   ____
   ____

2. The similarity in the early developmental stages of all vertebrate embryos suggests a common ancestral form.
   Neutral Report

   ____
   ____
   ____
   ____

3. In the Rococo period in Europe, architecture, sculpture, painting, and the decorative arts all exhibited a similar artistic style.
   Certain Report

   ____
   ____
   ____
   ____

4. The problem with personality typologies is that they tend to disregard situational variables.
   Neutral Report

   ____

   ____

   ____

   ____

5. Animal experimentation is not reliable because the underlying physiology of human beings is very different from that of lab animals.
   Certain Report

   ____

   ____

   ____

   ____

6. Material aspects of culture are more readily diffused than ideas or abstractions, and even then, it is rare for something to be borrowed intact.
   Doubtful Report

   ____

   ____

   ____

   ____

7. Craftsmen in colonial America used designs and techniques that were popular in England.
   Certain Report

   ____

   ____

   ____

   ____

8. The experimental design for many of the studies in parapsychology is flawed.
   Negative Report

   ___

   ___

   ___

   ___

9. About three billion years ago Mars changed from a hospitable planet to a frozen desert.
   Doubtful Report

   ___

   ___

   ___

   ___

10. Dancing began as part of the rituals that were practiced by ancient societies.
    Neutral Report

    ___

    ___

    ___

    ___

★**Bonus:** iBT Speaking
It is important to be able to report the information from a lecture as well as from a reading passage. Both of these prompts appear in the integrated tasks on the Speaking and Writing sections.

Writing Strategy 11

Use academic language

Prefer language that provides the possibility for another point of view. Avoid words like *absolutely*, *certainly*, *definitely*, and *obviously* because they are very assertive for academic writing. Statements like *I am sure* or *I am positive* may sound aggressive. *Always* and *never* invite the reader to find an exception. Use language that allows for an impartial exchange of ideas. Four main types of words and phrases for academic writing are listed below.

Verbs	**Modals**	**Adverbs**		**Adjectives**
Assume	Can	Apparently	As a rule	feasible
Believe	Could	Basically	All things considered	likely
Claim	May	Essentially	For the most part	possible
Suggest	Might	Generally	In general	plausible
Tend	Would	Mainly	On the whole	probable
Think		Primarily	Perhaps	
It appears that		Probably	Somewhat	
It seems that		Usually		

EXAMPLES

Assertive	**Academic**
Cell phones definitely cause most accidents.	It appears that cell phones cause most accidents.
Obviously, life is easier now than it was in the past.	Life seems easier now than it was in the past.
Class attendance must be required.	Class attendance might be required.
It is always important to be on time.	It is usually important to be on time.
I am positive that a large university will offer a superior education to that of a small college.	It is likely that a large university will offer a superior education to that of a small college.
I would never borrow money from a friend.	I would probably not borrow money from a friend.

Practice 11

First read the sentence and decide whether it is too assertive. If it is okay, mark it with a check mark. If it needs to be changed, use one of the academic options to rewrite the sentence. Then compare your answers with the example answers in the Answer Key on page 380.

1. It is obvious that everyone should attend college.

 ____

2. The university administration should never assign roommates to students.

 ____

3. I am sure that paying teachers better salaries would improve their performance.

 ____

4. If I could study in another country for one year, I would probably choose Canada.

 ____

5. Tests always serve a beneficial purpose in education.

 ____

6. I tend to prefer traditional classroom instruction to new technological options.

 ____

7. I am certain that cooperation is better than competition in the classroom.

 ____

8. A large university is definitely a better choice than a small college.

 ____

9. Teachers probably assign more homework than necessary.

 ____

10. For the most part, learning should be fun.

 ____

★**Bonus:** iBT Speaking
Assertive language is usually not appropriate in either academic speaking or academic writing.

Writing Strategy 12

Include a variety of sentence structures

You can add points to your score if you use different types of sentences in your essay, especially if you include compound and complex sentences. Even if your grammar is perfect, raters will not score your essay at the top range when all of the sentences are very simple and have the same structure. Be sure to consider variety and complexity when you write.

EXAMPLE ESSAY

According to the reading passage, sleep deprivation is common in our modern society. The amount of sleep required varies from person to person. One in five people fails to get enough sleep. Most people are working longer hours. Some people have two jobs. People lose sleep on a regular basis. They can suffer serious consequences. Relatively few studies have examined this important problem. The lecturer reported a research study. The research study was designed by graduate students at State University. They wanted to find out what happened to sleep-deprived college students. Researchers hypothesized that losing sleep would affect the students mentally and physically. The results confirmed the hypothesis. Sleep deprivation affected subjects in three ways. First, sleep deprivation affected their personalities. They became very irritable. For example, subjects with children lost patience more often. Small problems bothered them more than usual. Second, lack of sleep affected their cognitive processes. They made more mistakes. They performed tasks more slowly. In math tests, subjects received lower scores. They needed more time to complete them. Third, without enough sleep, subjects had physical problems. They were not as coordinated. Scores on computer games were lower. Scores were still lower with more time to finish the games. The immune systems of a significant number of subjects were depressed. They reported more sick days. They had more colds. In conclusion, getting enough sleep is important. It helps maintain good mental and physical health.

RATER COMMENTS: 3

This essay summarizes the reading passage. It is well organized, without grammatical errors. However, all of the sentences followed a similar pattern—simple sentence structure. This essay would have been excellent if the writer had used a variety of sentences, and had demonstrated the ability to use more complex sentences.

Example Essay Revised

According to the reading passage, sleep deprivation is common in our modern society. Although the amount of sleep required varies from person to person, one in five people fails to get enough sleep. Most people are working longer hours, and some people have two jobs. If people lose sleep on a regular basis, they can suffer serious consequences. However, relatively few studies have examined this important problem. The lecturer reported a research study, which was designed by graduate students at State University to find out what happened to sleep-deprived students. Results confirmed their hypothesis that losing sleep would affect the students both mentally and physically. Sleep deprivation affected subjects in three ways. First, sleep deprivation affected their personalities. For example, subjects became very irritable, those with children lost patience more often, and small problems bothered them more than usual. Second, lack of sleep affected their cognitive processes. They made more mistakes; moreover, they performed tasks more slowly. On math tests, subjects received lower scores, and they needed more time to complete them. Third, without enough sleep, subjects had physical problems. They were not as coordinated. Scores on computer games were lower even though subjects were provided with more time to finish them. Furthermore, the immune systems of a significant number of subjects were depressed, they reported more sick days, and they had more colds. Therefore, getting enough sleep is important because it helps maintain good mental and physical health.

Rater Comments: 4.5

This essay summarizes the reading passage. It is well organized, without grammatical errors. The writer used a variety of sentences, and demonstrated the ability to use more complex sentences.

Example Sentences

Series

Commas between parallel structures

For example, subjects became very irritable, those with children lost patience more often, and small problems bothered them more than usual.

Furthermore, the immune systems of a significant number of subjects were depressed, they reported more sick days, and they had more colds.

Relative Clauses

that, which

The lecturer reported a research study, which was designed by graduate students at State University to find out what happened to sleep-deprived students.

Compound Sentences

Addition *and*, Contrast *but*, Alternative *or*

Most people are working longer hours, and some people have two jobs.

On math tests, subjects received lower scores, and they needed more time to complete them.

Complex Sentences

Contrast: *although*, *though*, *even though*

Although the amount of sleep required varies from person to person, one in five people fails to get enough sleep.

Scores on computer games were lower, even though subjects were provided with more time to finish them.

Cause/result: *because, since*

Therefore, getting enough sleep is important because it helps maintain good mental and physical health.

Complex Sentences

Condition: *if*, *unless*

If people lose sleep on a regular basis, they can suffer serious consequences.

Conjoined Sentences

Addition: *moreover*; *furthermore*

They made more mistakes; moreover, they performed tasks more slowly.

Furthermore, the immune systems of a significant number of subjects were depressed, they reported more sick days, and they had more colds.

Contrast

however

However, relatively few studies have examined this important problem.

Conclusion

therefore

Therefore, getting enough sleep is important because it helps maintain good mental and physical health.

Practice 12

First, read the essay. Then read each set of simple sentences and combine them into one sentence. Using the cue that is provided, write a series, a relative clause, a compound, complex, or conjoined sentence. When you have completed all of the sets of sentences, compare your answers with the example answers in the Answer Key on page 380. Finally, read the revised essay.

According to the reading passage, Tyrannosaurs were among the most intimidating predators. Until recently, we assumed that they hunted alone. Canadian paleontologist Philip Currie has a new hypothesis. He thinks that these large dinosaurs hunted in groups. He calls these groups *dino gangs*. This idea is not exactly new. In excavation sites in Canada, many skeletons have been found in close proximity. Other sites in Argentina also uncovered the remains of groups. The Korea-Mongolia International Dinosaur Project provided more evidence. Six Tarbosaurus of different ages were found in close proximity in bone beds. Currie and associates interpret the skeletal remains as proof of social behavior. They assert that Tyrannosaurs lived and hunted together. In contrast, the lecturer disagrees with the theory. She proposes three alternative reasons why specimens were found together. The three reasons include natural disasters. The topography of the area is a reason. Famine is a reason. Natural disasters can cause animals to run to a location of safety. For example, a flood could cause the Tyrannosaurs to move to higher ground. The topography of the area could explain the presence of many skeletons. Soft mud or quicksand could trap animals. Famine could cause the population to congregate and eventually die in sparse feeding grounds. Philips could provide evidence of groups of footprints that appeared to be those of a group of hunters. The hypothesis would be more compelling. Pack hunting is a possibility. It is not the only interpretation of the evidence.

Conjoined Sentence
Contrast: *however*
Until recently, we assumed that they hunted alone.
Canadian paleontologist Philip Currie has a new hypothesis.

____

____

Relative Clauses: *that, which*
He thinks that these large dinosaurs hunted in groups.
He calls these groups *dino gangs*.

____

____

Conjoined Sentence
Addition: *moreover*
In excavation sites in Canada, many skeletons have been found in close proximity.
Other sites in Argentina also uncovered the remains of groups.

____

____

Compound Sentence
Addition: *and*
Currie and associates interpret the skeletal remains as proof of social behavior.
They assert that Tyrannosaurs lived and hunted together.

Series: Commas
She proposes three alternative reasons why specimens were found together.
The three reasons include natural disasters.
The topography of the area is a reason.
Famine is a reason.

Complex Sentence
Cause/result: *because*
The topography of the area could explain the presence of many skeletons.
Soft mud or quicksand could trap animals.

Complex Sentence
Condition: *if*
Philips could provide evidence of groups of footprints that appeared to be those of a group of hunters.
The hypothesis would be more compelling.

Compound Sentence
Contrast: *but*
Pack hunting is a possibility.
It is not the only interpretation of the evidence.

★**Bonus:** iBT Speaking
A variety of sentence structures is also impressive on Tasks 4 and 6 in the Speaking section because you are asked to summarize and synthesize formal academic lectures.

Writing Strategy 13

Follow a plan for the integrated essay

You need a plan to organize the integrated essay. First, summarize the reading passage. Then use a transition sentence to signal whether the lecture supports (agrees with) or contradicts (disagrees with) the ideas in the reading passage. Next, summarize the lecture. Finally, show how the lecture and the reading relate. Keep these words in mind as you organize your essay: summarize, transition, summarize, and relate. Transitions that signal support or contradiction are listed below.

Transitions that signal support	***Transitions that signal contradiction***
The lecture *confirms* the information in the reading.	The lecture *contradicts* the information in the reading.
In support of the information in the reading, the lecturer . . .	In contrast with the information in the reading, the lecturer . . .

Example Reading Passage

Organically grown food, which contains no additives and is produced without chemical fertilizer and pesticides, has become progressively more popular in Britain, Europe, and the United States. To determine whether the nutritional value of organic food justifies the higher cost for consumers, a major research investigation was commissioned by Britain's Food Standards Agency, the FSA, and conducted by a team of researchers from the London School for Hygiene and Tropical Health. Published in the *American Journal of Clinical Nutrition,* the study reviewed and summarized scientific papers published in the past 50 years on the benefits of organic food to health and diet.

Researchers determined that the small number of differences in nutritional content found in organically grown food was unlikely to affect the health of consumers. Both organic and conventional food were comparable in nutritional values of vitamin C, calcium, iron, and fatty acids. In short, the team concluded that there was no evidence to encourage the selection of organically grown food in preference to conventionally grown food on the basis of nutrition. Although more phosphorus was identified in organically grown food, it was dismissed as irrelevant to public health because sufficient levels of phosphorus are also commonly found in most conventionally grown food. Levels of beta-carotene, which were identified at levels as much as 53 percent higher in organic food, were listed in the appendix but did not figure in the findings.

During the investigation, it was also found that acidity was slightly higher in organic produce; however, acidity was determined to affect taste and perception rather than nutritional values, according to the study. No evidence was cited to prove that additional health benefits were associated with eating more acidic organic food.

EXAMPLE LECTURE SCRIPT

In response to the recent study by the FSA in Britain, which suggested that the nutritional values of organic food did not differ significantly from that of conventionally grown food, several research studies in the United States and Europe have released contradictory findings. For example, in a study of strawberries, University of California researchers found that the antioxidant content was significantly higher in the organic fruit as compared with conventionally grown fruit. This study confirmed the findings of AFSSA, the French Agency for Food Safety, which identified higher mineral content and antioxidant micronutrients in organic food than in their conventional counterparts. Recent reports published jointly by the University of Florida and Washington State University suggested that organically grown food contained an average of 25 percent higher concentrations of 11 important nutrients, and researchers at the University of California at Davis also identified higher levels of beneficial flavinoids in tomatoes.

In addition, it was not mentioned in the British study that organic food offers more varieties from which consumers can choose. In side-by-side studies in the United States, results suggested that organic fruit and vegetables were consistently judged to be more visually appealing and more flavorful, and thus, consumers were more likely to choose organic fruit and vegetables instead of eating less nutritious food choices.

Furthermore, it should be pointed out that investigators for the British study did not consider pesticide and herbicide residues in their comparisons. In conventional farming, food can be sprayed as many as 16 times with 30 different pesticides. In almost every study in the United States in which pesticides and herbicides were measured, significantly lower residues of toxic chemicals were found in organic food as compared with conventionally grown food. Perhaps even more important was a study at the University of Washington comparing pesticide metabolites in children who ate an organic diet and those who ate a conventional diet. Children tested in the conventional diet group showed six times the amount of pesticide metabolites as the organic diet group, bringing their exposure to levels above the U.S. Environmental Protection Agency's current guidelines.

Example of Integrated Essay

Summarize:

To establish whether organically grown food provided enough nutritional benefit to warrant the additional cost to consumers, Britain's food standards agency, the FSA, commissioned a research team from the London School for Hygiene and Tropical Health to review studies on the advantages of organic food published in scientific journals over the past 50 years. Investigators concluded that the differences between the nutritional values in organic food and conventionally grown food were too small to affect the health of consumers and therefore, on the basis of nutrition, organic food was not worth the higher prices.

Transtition:

In contrast with the information in the reading, the lecturer reported the findings from a number of studies at American and French universities confirming higher nutritional values in organically grown food.

Summarize:

Specifically, antioxidant levels were higher in organically grown fruit, mineral content and antioxidant micronutrients were higher in organic food, and flavonoids were higher in tomatoes. One study identified an average of 25 percent higher levels of 11 nutrients. Researchers also criticized the British study for failing to discuss the advantage of more visual appeal, additional varieties, and better flavor, all of which would encourage consumers to eat organic fruit and vegetables instead of choosing food that may not be as nutritious. In addition, pesticide and herbicide residues were not measured in the British study. In American studies, however, higher toxic chemical residues were found in both conventional foods and in children who ate a conventional diet as compared with organic foods and children who ate an organic diet.

Relate:

Because of the contradictory findings of the British, French, and American studies, further study of the issue is probably justified.

Practice 13

First, read the passage. Then listen to the lecture. Finally, use the plan to write a synthesis about 250 words in length. Compare your answer with the synthesis in the Answer Key on page 382.

Reading Passage

Until recently, most archeologists concurred with the theory that the original American Indians migrated from Asia across the Bering Straits about 12,000 years ago when the two continents were connected by a frozen land bridge. According to the Beringia theory, migrants were following herds of big game that continually pressed south in search of grazing lands in warmer climates. Bands of nomadic hunters followed the herds through a green corridor between glaciers along the Rocky Mountain foothills, into the Great Plains, and across the American Southwest.

The path was marked by flint arrowheads and spear points that were first discovered in Clovis, New Mexico, and later excavated in a large number of sites north and south of the original find. The projectiles were crafted with a flaked flint point into which a notched flute was cut. This distinctive flute identified and united the so-called Clovis culture. By mapping the locations where Clovis points were discovered, it appeared that the Clovis people had traveled as far as the tip of South America, establishing tribes of Indians throughout the New World. Perhaps even more astonishing was evidence from carbon dating, which suggested that the migration from Siberia to Tierra del Fuego had been accomplished in about 1,000 years.

A number of newer theories have been put forward to challenge the Beringia theory, including several alternative routes that allowed for water migrations, either across the ocean from Europe along the edges of ice flows and glaciers in the North Atlantic or across the Pacific and down the west coasts of North and South America. Recent excavations, including the Monte Verde site in Chile and several sites in South Carolina and Virginia have provided archeological evidence that calls into question the Clovis origins. According to several noted anthropologists, some of the sites could predate the Clovis culture. Although the new excavations do not necessarily support any specific water migration theory, it does appear that the paradigm for migration patterns in the New World may be on the brink of change.

Lecture

CD 2, Track 16

Synthesis

Summarize:

Transition:

Summarize:

Relate:

★**Bonus:** iBT Speaking

A reading and a lecture appear in Task 4 of the Speaking section on the iBT as well as in the integrated essay on the iBT. You can use the same plan to speak or write.

Writing Strategy 14

Use a pattern for the independent essay

Plan your independent essay using a traditional pattern of organization.

Paragraph 1:	Optional introduction with thesis statement Outline sentence with three ideas
Paragraph 2:	Idea 1
Paragraph 3:	Idea 2
Paragraph 4:	Idea 3
Paragraph 5:	Conclusion

You can see why a short essay in English is often called a five-paragraph essay. By following this plan, you will develop your essay in the order that raters expect, and you will be more likely to receive a higher score for organization.

Example of Independent Essay

Paragraph 1: Optional introduction

Although international students face a number of problems when they decide to study at a university in an English-speaking country, in my view, the most challenging is mastering the English language.

Outline Sentence

<u>This is a formidable task because the level of mastery must be high enough for students to meet requirements for admission, to understand how to cope in a different culture, and to compete with native speakers in classrooms</u>.

Paragraph 2: Idea 1

<u>Most major institutions require some type of language proficiency test as part of the application and admission process for international students</u>. One of the most widely accepted tests is the TOEFL, which requires that students perform well on Reading, Listening, Speaking, and Writing sections. Without the required TOEFL scores, international students may not be accepted to one of the universities where they have applied. Therefore, the first challenge is to learn enough English to achieve a high score on the test.

Paragraph 3: Idea 2

<u>After being admitted to a university and arriving on campus, international students must use their English language skills to solve all of their other problems, including where to live, how to find and use basic services such as grocery stores, pharmacies, laundries, and public transportation</u>. With a command of English that supports an understanding of the culture and the social skills to interact with other students, professors, and campus personnel, international students will be able to cope in a new and very different environment, and they will be able to solicit the assistance of people who can provide resources and advice. Once again, mastery of English is crucial.

Paragraph 4: Idea 3

In a university classroom, competition is assumed. Many professors grade on a curve, which means that all students are measured not only in terms of their actual performance but also in comparison with the performance of all other students in the class. Even when other, more objective standards are used, students are constantly being judged against their classmates. For international students, this means that they must learn the content of the class and also learn English at a level that will allow them to express their knowledge in academic language on tests, papers, projects, and other assignments.

Paragraph 5: Conclusion

In summary, with a high degree of competence in the English language, international students will be able to solve many of the other problems that they encounter. Specifically, they will succeed on the tests required for them to gain admission to the universities that they want to attend, they will be able to adjust and cope with the very different cultural environment, and they will be able to compete with other students in their classes.

Practice 14

First, read the essay. Then identify each part and label it, using the labels on page 271. Finally, check your answers in the Answer Key on page 383.

Independent Essay

I recognize that some people would prefer to use the investment in space exploration for immediate improvements to life on Earth, but I believe that it is important to continue studying the universe because exploration may provide the best long-term solutions to some of the most serious problems that we face on our planet. While I could discuss a number of these problems, the pollution of the environment, the depletion of natural resources, and overpopulation come to mind.

First, it is important to acknowledge that while scientists are exploring space, serious environmental studies of Earth are also being conducted. For example, when scientists travel into space, they also collect samples of air to determine current air quality in the atmosphere and to track changes in our most necessary requirement for life on Earth—the air we breathe. Furthermore, information about climate change can best be examined from space. The movement of bodies of water, the thickness of ice, and the flow of jet streams and weather patterns are all studied during space exploration.

In addition, exploring other worlds could result in the discovery of valuable natural resources. These resources may include many energy sources that are currently being depleted on Earth or even presently unknown resources that could lead to a total revolution in the way that we design our living spaces, vehicles, and technologies. Few would disagree that we are consuming the resources that we have available on Earth. If we do not explore extraterrestrial alternatives, eventually, we will run out of options.

Finally, it is obvious that the population on Earth is continuing to increase. Almost 7 billion people now share the planet, and some argue that the current figures already represent a strain on the number that our planet can support. If we could locate a planet or moon that would support life as we know it, colonization would be an obvious solution to overpopulation. We should not wait for an emergency to explore new places to settle.

Therefore, although funding for the infrastructure and social programs on Earth may seem more attractive and would certainly show a more immediate return on investment, space exploration offers not only the chance to study outer space but also the opportunity to find better solutions to problems on Earth.

LABELS FOR INDEPENDENT ESSAY

Optional introduction

Outline sentence

Idea 1

Idea 2

Idea 3

Conclusion

Writing Strategy 15

Announce your conclusion

Your conclusion will leave the rater with a final impression and can make the difference between a 4 and a 5 on your essay. You can use the conclusion to summarize your essay or bring all the points together. To let the reader know that you are concluding your essay, choose from the words and phrases listed below.

To conclude,
To recap,
To recapitulate,
To reiterate,
To summarize,
To sum up,
In conclusion,
In short,
In summary,
In closing,

Examples of Conclusions

In conclusion, the perfect job would include a good salary, benefits, and opportunities for advancement.

To summarize, trade unions demanded shorter working days, better working conditions, and the end of child labor.

To recapitulate, the automobile was probably the most significant invention in modern times because it improved transportation, provided a boost for the economy, and introduced mass production as a model for other products.

In closing, I believe that living in a dormitory offers more advantages than disadvantages for international students.

To reiterate, good luck is important, but hard work is probably even more crucial to long-term success.

Practice 15

First, read a set of three related sentences. Each set includes the first sentence for each paragraph in the body of an essay. Then, write a conclusion that summarizes the sentences or relates the ideas. Compare your answers with the example answers in the Answer Key on page 384.

Agree or disagree with the following statement: States should not use lotteries to raise money.

1. In the first place, the odds of winning a lottery are very slight.
2. In addition, many people who cannot afford to spend money on the lottery will buy tickets.
3. Finally, gambling can become addictive.

Conclusion:

____

____

Based on the reading passage and the lecture, was Shakespeare the author of the plays attributed to him?

1. According to the reading, the historical record shows Shakespeare's name on the plays and on related documents in reference to the plays.
2. Furthermore, contemporaries of Shakespeare did not doubt the authorship.
3. The lecturer points out that Shakespeare was part owner in the Globe Theater and also a principal actor in the company.

Conclusion:

____

____

Agree or disagree with the following statement: College students should be required to take physical education classes.

1. The fees for physical education classes are the same as those for important academic subjects.
2. The time that is spent in physical education classes could be better spent studying for tests and preparing papers.
3. College students get enough exercise by walking or biking to class or by participating in recreational activities.

Conclusion:

____

____

What are the characteristics of good leaders?

1. Perhaps the most important characteristic of good leaders is the ability to surround themselves with talented people.
2. Good organizational and management skills are also essential to leadership.
3. Last, but equally important, good leaders must possess the ability to inspire others to action.

Conclusion:

____

____

Based on studies reported in the reading passage and the lecture, bilinguals demonstrated certain advantages because of an efficient executive control system in the brain's network.

1. In the first place, bilingual subjects tended to multitask more easily than monolingual subjects, and they made fewer errors while attending to the tasks.
2. Moreover, bilingual subjects demonstrated superior problem-solving skills when presented with both verbal and non-verbal problems.
3. Creativity, as measured by divergent thinking tasks, was also an area in which bilinguals excelled.

Conclusion:

____

____

Writing Strategy 16

Budget your time for the writing process

Be sure to take time to plan your essay before you begin, and save time to edit it after you have finished writing. The two most common mistakes that writers make on an essay test is that they begin to write before they have planned the essay and they don't save time at the end to correct errors that they could find if they reviewed their work. Research indicates that if you don't reread your essay, you will lose points. Use the clock to budget your time for the three parts of the writing process—planning, writing, and editing.

Integrated Essay		**Independent Essay**	
After Reading Passage		***No Reading Passage***	
20 minutes		30 minutes	
Planning	4 minutes	Planning	5 minutes
Writing	12 minutes	Writing	20 minutes
Editing	4 minutes	Editing	5 minutes

Example Clock for Integrated Essay

The clock starts at 20 minutes and counts down to 0.

20:00	***Planning*** Use a map, an outline, or organized notes.
16:00	***Writing*** Use the strategies in this chapter to improve your skills.
04:00	***Editing*** Use the Checklists in Strategy 19.

EXAMPLE CLOCK FOR INDEPENDENT ESSAY

The clock starts at 30 minutes and counts down to 0.

30:00	***Planning*** Use a map, an outline, or organized notes.
25:00	***Writing*** Use the strategies in this chapter to improve your skills.
05:00	***Editing*** Use the Checklists in Strategy 19.

EXAMPLES OF PLANNING OPTIONS

I think that physical education is an important part of an educational program.

Map

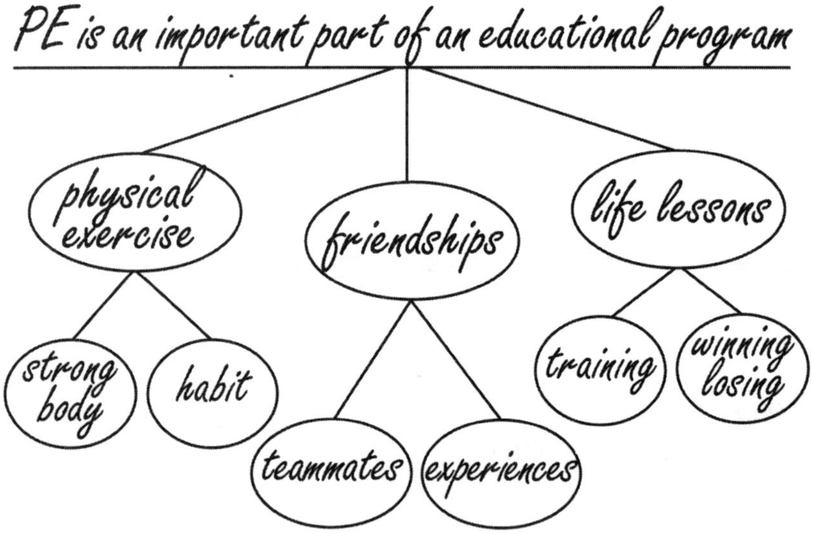

Outline

Physical exercise
- Strong body
- Habit

Friendships
- Teammates
- Experiences

Life lessons
- Training
- Winning and losing

Notes

Physical exercise
Friendships
Life lessons

Practice 16

First read the question. Then set a timer for 5 minutes and plan your essay. Try maps, outlines, lists, and notes to determine which method works best for you. Next, set the timer for 20 minutes and write your essay. Then, set the timer for 5 minutes and read your essay. Try to find and correct errors. By repeating this process ten times for different topics, you will start to develop a sense of the timing on the independent essay. Finally, compare your answers with those in the Answer Key on page 386.

1. Agree or disagree with the following statement: Everyone should attend college.

2. Agree or disagree with the following statement: The university administration should assign roommates to students.

3. Agree or disagree with the following statement: Paying teachers better salaries would improve their performance.

4. If you could study in another country for one year, where would you choose to go?

5. Some people believe that standardized tests serve a beneficial purpose in education. Other people think that they do not provide an accurate indication of a student's abilities. What do you think and why?

6. Some students prefer traditional classroom instruction. Others would rather study independently, using new technological options. Which method of instruction would you prefer, and why?

7. Agree or disagree with the following statement: Cooperation is better than competition in the classroom.

8. If you could choose between a large university or a small college, which would you prefer?

9. Agree or disagree with the following statement: Teachers assign more homework than necessary.

10. Some people think that learning should be fun. Other people believe that it is not important to have fun in the classroom. What is your opinion and why?

Writing Strategy 17

Choose verb tenses consistently

Be consistent in using verb tenses. Use present tenses to discuss scientific facts. Use past tenses to summarize a reading passage or lecture that you have read or heard in the past. Make sure that you use the same tenses throughout. Examples of present and past tense forms are listed below.

Present Tenses		***Past Tenses***	
-s	appears to change	*-ed*	changed
present BE + *-ing*	is changing	past BE + *-ing*	was changing
modal + verb	may change	HAVE + participle	had changed
present BE + *-ed* (passive)	is changed	past BE + *-ed* (passive)	was changed

EXAMPLE READING PASSSAGE *PRESENT TENSE*

Many educational researchers in the United States maintain that students who participate in bilingual education programs learn to read and speak English more rapidly and at higher levels of proficiency than students who are enrolled in English immersion programs that restrict or prohibit the use of the native language. According to the Interdependence Hypothesis put forward by Stephen Krashen and James Cummins, students learn to read once and then transfer reading skills from one language to another as they acquire a second or even a third language. Furthermore, they argue that it is easier and faster for these students to learn how to read in the native language.

By making progress in native language literacy, students are able to read content at levels that challenge them intellectually and they are more successful in all of their school subjects. Research studies by Krashen and others indicate that a child's proficiency in reading in the native language is a strong predictor of that child's performance in reading English and in succeeding in other subject areas in English. Therefore, bilingual education promotes literacy, English language proficiency, and content learning.

Robert Slavin and Alan Cheung recommend native language literacy programs; however, they also admonish that the most important factor is not the type of program but rather the quality of the program. According to their studies, English language learners benefit most from programs that include structure, cooperative learning strategies, tutoring either in small groups or individually, and systematic instruction from one grade to the next.

EXAMPLE LECTURE SCRIPT *PAST TENSE*

Much of the empirical evidence for the Interdependence Hypothesis has been derived from case studies of children who learned to read well in the first language, and subsequently acquired high levels of reading proficiency in the second language. In a longitudinal study by Krashen, Dow, and Tinajero, the hypothesis was tested in the El Paso schools. Investigators hypothesized that first language reading ability in grade 2 would correlate with second language reading ability in later grades. To test the hypothesis, 109 children were tested in Spanish on the Aprenda exam in grade 2, and in English on the SAT 10 four years later in grade 6. The correlation between reading in Spanish in grade 2 and reading in English in grade 6 was positive and statistically significant, confirming that those children who learned to read at high levels in early grades in their home language would read at higher levels in their second language.

In addition to the process of reading which underlies both languages, researchers assumed that children who learned to read well early in their school experience had more access to school subjects and acquired greater background knowledge that supported their learning in later grades.

Although some educators have used the results to make a case for bilingual education, others pointed to first language literacy as the key to success for second language learners, whether it was included in a bilingual education program or included as a component of a second language immersion program. Apart from the debate on programming, the study provided strong evidence for the efficacy of the Interdependence Hypothesis.

EXAMPLE SYNTHESIS

According to the Interdependence Hypothesis proposed by Krashen and Cummins and reported in the reading passage, it was easier and quicker for students to learn to read in their first language. Later, if they learned a second language, they could transfer all the reading skills that they had acquired when they learned to read initially. Furthermore, the reading proficiency demonstrated in the first language was an indicator not only of success in reading in English as a second language but also of academic success in other content subjects. Therefore, first language literacy was cited as one of the most important components of programs for English language learners, along with structure and coordinated instruction among the grades.

Although case studies traditionally have provided support for the hypothesis, the lecturer cited a longitudinal study in the El Paso school district that added significant evidence. A high positive correlation between reading proficiency in Spanish in the second grade and reading proficiency in English in the sixth grade demonstrated that the reading process acquired in the first language transferred to the second language. In addition, it was also suggested that early literacy allowed students to acquire higher levels of background knowledge in school subjects, an advantage to learning in upper grades.

The results of the El Paso study have been used to promote first language literacy in both bilingual and ESL programs. The important point was that the study made a strong case for the Interdependence Hypothesis.

Practice 17

First, read the synthesis and underline all of the verbs. Then draw a box around ten verbs that are not consistent with the tenses used in the summary. Finally, correct the verbs. Check your answers in the Answer Key on page 388.

Synthesis

The reading passage described a classic psychology experiment. Philip Zimbardo and colleagues at Stanford University designed a study to determine how participants would respond to a simulation of a prison environment. Volunteers for the study are white middle-class males who had no criminal record and no psychological issues. The question was whether good people would retain their normal reactions when placed in a prison environment.

The simulated prison consists of three cells, six by nine feet, with three cots in each cell to accommodate three prisoners. Other rooms across from the cells were designated for the prison guards and the warden. The volunteers were randomly selected to participate as either a prisoner or a guard. Prisoners were arrested, searched, fingerprinted, instructed to pose for mug shots, dressed in typical prison smocks with numbers on them and a stocking cap. They were also chained around their ankles. Guards were dressed in khaki uniforms and were provided with sunglasses to eliminate eye contact with the prisoners. They carry wooden batons, which identifies their authority, but were not to be used to punish prisoners. Guards were to identify each prisoner by his number, not by name. Prisoners are confined to their cells for the duration of the study, whereas guards were assigned to eight-hour shifts, and were free to return home between shifts. Researchers observe and record behavior using cameras and microphones concealed in the experimental areas.

The lecturer reported that guards were free to interact with the prisoners as they deemed appropriate, but many of them become aggressive or even abusive toward the prisoners. After a riot on the first day to protest the stocking caps that they were forced to wear, most of the prisoners became withdrawn and passive. Although the experiment was originally designed for 14 days, five prisoner participants left the experiment early, one research assistant objected to the abusive situation, and the entire experiment is discontinued after only 6 days because the remaining prisoners were beginning to demonstrate extreme stress and anxiety. In conclusion, the Stanford Prison Experiment demonstrated the very powerful influence of situations in human behavior. Only a few people are able to behave according to their usual moral dictates.

★**Bonus:** iBT Speaking

Consistent verb tenses are not as noticeable in speaking as they are in writing, but they impress raters when they are used correctly on the Speaking section.

Writing Strategy 18

Review your essays for basic capitalization and punctuation

You cannot access Spellcheck and Grammar Checker in the writing program on the TOEFL. Although most raters will not deduct points for minor errors, some raters are influenced by basic mistakes because they may be an indication of carelessness or beginning levels of English proficiency. Be sure to reread your writing and correct small mistakes in capitalization and punctuation before you submit your essays.

EXAMPLES

Indentation

Indent every paragraph. It is also correct to leave space between paragraphs instead of indenting, but the space can sometimes close when an essay is saved. It is better to indent.

[➔] Although the English language does not have a direct correspondence between the sounds and the spelling of words, some basic spelling rules are taught in schools.

Period

Use a period at the end of every sentence. A period is called a *full stop* in British English.

Children often learn phonics, which is a system of sounding out words [.]

Comma

Use a comma to separate the items in a series.

Short vowels [,] long vowels [,] and the schwa vowel must be taught so that children can connect the sounds with the letters on a page or a screen.

Colon

Use a colon before a series. A colon is often but not always introduced by the phrase *the following.*

When a whole language approach was introduced in the 1980s, phonics was replaced by the following strategies [:] opportunities to read interesting literature, motivation to become good readers, and focus on meaning instead of sounds.

Capital Letters

Use a capital letter at the beginning of the first word in a sentence.

[N]ow phonics is being reintroduced in many schools to supplement whole language programs.

Practice 18

Review the practice essay for basic capitalization and punctuation rules. Look for ten errors, and check your answers in the Answer Key on page 389.

Although some students prefer a traditional classroom, I like to take classes online. These nontraditional classes offer the following advantages greater flexibility increased communication, and lower cost.

Perhaps the most important attraction for me personally is that an online class schedule can be adjusted to fit my routine. Since I work full time, traditional courses are often offered when I am obligated to be on the job In addition, I am married with two children, and my family deserves my attention and attendance at school events. the time that I would spend driving to campus and finding a parking space can be used for work and family activities. It is not uncommon for me to study at night after I have finished work and the children are in bed.

Another advantage for online instruction is the opportunity to communicate more effectively with the teacher and other students. I tend to be rather shy about speaking in class, especially in a large lecture setting.

Online classes offer me the opportunity to ask questions by e-mail. I also notice that I interact more with my classmates because chat rooms and threads online support communication. In a traditional class, only a few students tend to talk, but online more of us are asking questions and responding to each other

Tuition for an online class varies with the school and degree program in much the same way that it does for a traditional class, but in my experience, the cost of books and lab materials is lower for online classes. many programs do not require textbooks, which tend to be very expensive. On the contrary, the materials for classes and labs are often available as part of the fees for online instruction and can be downloaded along with lectures handouts, additional notes, and other resources.

I do not claim that online classes should replace traditional classes for everyone. In fact, a very good case could be made for them at certain stages of life. What I maintain is that for me, at this point in my life, online classes offer the better opportunity for me to study while I am meeting my work obligations and my family responsibilities

Writing Strategy 19

Use the rater's checklists to correct your essays

Take time to read and understand the rater's checklists so that you can have them in mind when you write and edit your essays. In general, raters are trained to ignore minor errors because they do not expect non-native writers to produce perfect essays. If an error interferes with comprehension or changes the meaning of the reading passage or lecture referred to in the essay, the rater will judge it a major error and you will lose points, but if an error does not cause the reader to stop and think in order to interpret the meaning, the rater will judge it a minor error, and will probably not lower your score. Leave time at the end of your essays to edit your writing. Learn to think like a rater.

Example of Rater's Checklist: 5

✔ The writer responds to the topic and the task.
✔ The essay is well organized and well developed.
✔ The sentences are logically connected by transitions.
✔ The content is accurate in the integrated essay.
✔ Details and examples support the topic sentence.
✔ The meaning is easy to comprehend.
✔ A wide range of appropriate vocabulary is used.
✔ Sentence structure is varied at an advanced level.
✔ Errors in grammar are minor or repeated.
✔ The independent essay is within a range of 300–350 words.
✔ The integrated essay is within a range of 150–225 words.

Example of Rater's Checklist: 4

✔ The writer responds to the topic and the task.
✔ The essay is somewhat organized but not fully developed.
✔ The sentences are usually logically connected by transitions.
✔ The content is accurate but may be incomplete in the integrated essay.
✔ Details and examples do not fully support the topic sentence.
✔ The meaning is mostly comprehensible.
✔ A good range of appropriate vocabulary is used.
✔ Sentence structure is varied at an intermediate level.
✔ Errors in grammar are noticeable but do not interfere with meaning.
✔ The independent essay is within a range of 300–350 words.
✔ The integrated essay is within a range of 150–225 words.

EXAMPLE OF RATER'S CHECKLIST: 3

- ✔ The writer responds to the topic and the task.
- ✔ The essay is somewhat organized but only partially developed.
- ✔ The sentences are usually logically connected but transitions may be unclear.
- ✔ The content is not always accurate and is incomplete in the integrated essay.
- ✔ A few details and examples support the topic sentence.
- ✔ The meaning requires thought to comprehend.
- ✔ A limited range of appropriate vocabulary is used.
- ✔ Sentence structure is limited.
- ✔ Errors in grammar result in lack of clarity.
- ✔ The independent essay may not be within a range of 300–350 words.
- ✔ The integrated essay may not be within a range of 150–225 words.

Practice 19

First read the essays. Next refer to the checklists to evaluate the writers. How would you change the essays? Compare your evaluations with the evaluations in the Answer Key on page 390.

Independent Essay 1

When I have an important decision to make, I usually follow a three-step plan. First, I do research, then I make a list of the advantages and disadvantages for each choice, and finally I talk about it with my family. One example that comes to mind is my decision to enroll in a masters degree program in education at Columbia University.

When I was applying to graduate programs, I used catalogues and the Internet to locate the best programs in my field. I eliminated programs that were not accredited, and I applied to four programs that had excellent ratings in education. Although I had a very good undergraduate record, my research indicated that I should probably apply to at least one program that was less competitive. In addition to the Ivy League schools that attracted me, I also applied to a state university in New York. Because of the fees and the time involved to complete each application package, I decided to focus on five schools.

As I began to make lists of the advantages and disadvantages for each school, I considered the program, the location, the cost, and the possibility of receiving a grant or a teaching assistantship to help me fund my education. It was fairly easy to make a chart, and by comparing all of the criteria, I was able to put the schools in order from 1 to 5, with 1 as the most desirable and 5 as the least desirable. In the end, I think that my emotional response to the schools also contributed somewhat to the numerical order, even though I did not include it in the chart.

Of course, I was talking with my brother and my parents while I was doing the research and evaluating the advantages and disadvantages of the schools; however, we did not have our family meeting until I began to receive responses from my applications. Within a two-week period, I received responses from four of the five schools. I was accepted at three of them and put on a wait list for a fourth, but I probably would not know whether I would be accepted there until very close to the beginning of the school year.

My family and I had our meeting in the dining room around the table. Since this room is used for formal events, we all realized that this was an important meeting. I told my family that I had been accepted by two of my top three schools and that I would be happy to attend either one of them. Unfortunately, I did not receive financial aid from any of the schools. We talked about the advantages and disadvantages of each school and everyone expressed an opinion.

In conclusion, the three-step plan proved to be a very useful tool for making this important decision. In the end, I decided to attend Columbia because my uncle lives in New York, the fees were slightly cheaper, and it would be easier to visit my family because direct flights from my city to New York are often discounted. I was prepared to attend any of the five schools, but the program at Columbia was first on my list from the beginning. I have used the three-step plan since then to help me negotiate other important decisions and I know that it will be influential in my thinking for the rest of my life.

Level: ___________

Rater's Comments

____

____

____

____

____

Independent Essay 2

When I have an important decision, I use three steps to help me. First, I researched the situation. Next, I make a list of the advantages beside a list of the disadvantages for each choice. Finally, I talk it over with my family. That is what I did when I decided to go to Columbia University for my master degree.

I used Internet and some catalogues to find the best schools in my field before I applied to them. My undergraduate record was very strong but I wanted attend an ivy league school so I applied to four of them, but I also applied to a state school that was less competition to get accepting to it as backup. The expensive of the fees for applying made me stop at five schools.

I was thinking about the place where is the school, how much did it cost, how good is the program, and financial aiding so I made a chart of all of these things and look at it from 1 to 5 of my choice. I was also thinking about the schools that I liked best but I was trying to use advantages only for decision.

I got accepting at three schools with waiting at one school before my family meeting but I did not get financial aid at the schools. I told my family about my accepting at my two best schools and everyone said their opinions about it. The program at Columbia was first choice for me and we decided it because my uncel lives in New York and flying to my country is easy, and fees are more cheaper.

This three steps are good for me to make big decisions and I will use this plan again and again in my life.

Level: __________

Rater's Comments

____

____

____

____

____

Integrated Essay 1

Professor Chandra Wickramasinghe and Sir Fred Hoyle of Great Britain put forward a claim that comets may be responsible for some of the diseases that have afflicted the Earth over the centuries, including plagues and major flu epidemics. According to their theory, dust particles that fill the space between stars consist of tiny living organisms. Moreover, since comets are comprised of interstellar dust blended with ice and frozen gasses, they are natural carriers of the organisms as they complete their orbits. In other words, when a comet passes close to Earth, some of its dust enters our planetary atmosphere. There, the minute organisms multiply and drift into the lower atmosphere where humans can become exposed to them.

Hoyle and Wickramasinghe cite several examples of diseases that could possibly be attributed to exposure from the dust carried by comets, and in some cases, even identify a specific comet that may have been the carrier of a particular disease. For example, Halley's comet orbits the Sun in 75 to 78-year cycles. In 1957, a pandemic of Asian flu occurred, and 77 years later, another global outbreak coincided with the comet's orbit. According to the theory, these outbreaks of flu may have been caused by dust from the comet. In addition, major smallpox epidemics seem to have occurred in regular cycles of about two hundred years, which Wickramasinghe and Hoyle believe may coincide with an undiscovered comet.

To test their theory, WIckramasinghe and Hoyle studied the spread of flu in British boarding schools. Instead of spreading from one dormitory to the next, the flu appeared randomly at the same time in different dormitories. They concluded that the flu may have been caused by organisms in the atmosphere. On a wider scale, they looked at the spread of flu in pandemics prior to the introduction of air travel. They observed that although person-to-person infection occurred in local areas, the diseases also appeared on the same day in widely separated areas of the world.

Level: __________

Rater's Comments

____

____

____

____

____

Integrated Essay 2

In the reading passage, the theory that comets responsible for diseases on Earth like plagues and flu epidemics. They say that dust particles fill up the spaces between stars and they have little organisms so the comets are ice, gas, and dust that can carry the organisms with them on the obits. Therefore, comets comes close to Earth and then the dust comes down to the atmosphere and humans get exposed. So some diseases are Asian flu and smallpox that happen when comets orbiting like Halley's Comet or another one.

The lecture talked about how to test its theory. Hoyle and Wickramasingh did a research at British boarding schools about how did flu spreading. Instead spreading one dormitory beside next, flu came random dormitories for the same time. So conclusion that flu is happening by organisms up the atmosphere instead passing from people. Pandemics also showed infections for the same time at farther places of the world. This is proving the theory.

Level: __________

Rater's Comments

____

____

____

____

____

Writing Strategy 20

Train mentally and physically for a successful finish

The TOEFL is a very long test and the Writing section is presented at the end when you may be getting tired. You will need to train mentally and physically, just like a marathon runner, in order to finish strong. To train mentally, visualize your goal. When the screen for the Writing section appears, see yourself receiving an acceptance letter to the school of your choice or obtaining your license, a scholarship, or a degree. Don't get lost daydreaming. Just see it in a flash. Take a deep breath and get ready to push harder to the finish line. To train physically, take full-length practice tests without stopping. Use all of the strategies that you have learned and keep going until you finish the last question. Your training will pay off when you take the official TOEFL, and you will win the race!

Example Resources for Full-Length Practice Tests

There are many resources for you to use when you train for the TOEFL. When you make a choice, be certain that all of the sections are included in the practice test and that the time for the practice test is the same as that of the official TOEFL. The following packages are carefully designed to provide a very similar experience. They can be located in bookstores worldwide, in Internet bookstores, and through the publisher's website at *www.barronseduc.com.*

Barron's TOEFL iBT 13th Edition with CD-ROM
Dr. Pamela J. Sharpe
Barron's Educational Series
ISBN 978-0-7641-9698-0

CD-ROM includes eight full-length practice tests that simulate the 4-hour iBT TOEFL.

Book includes seven of the full-length practice tests in written format.

This is the best option for practice, using full-length tests.

Barron's Practice Exercises for the TOEFL 7th Edition with Audio CDs
Dr. Pamela J. Sharpe
Barron's Educational Series
ISBN 978-1-4380-7033-9

Audio CDs include one full-length practice test that simulates the 3-hour ITP TOEFL and one full-length practice test that simulates the 4-hour iBT TOEFL. Computer visuals do not appear on audio CDs.

Book includes 100 practice exercises in addition to the tests.

This is an alternative for practice if you are taking either the iBT or the ITP test.

Practice 20

First think about your goal. Why are you taking the TOEFL? What will a successful score do for you in the short run and in the long run? Try to visualize yourself achieving your goal. Learn how to bring up that image quickly. Then select a resource for full-length practice tests. When you practice using one of the full-length tests, be sure to complete all of the sections without stopping. Visualize your image when the Writing section begins. Your training with full-length practice tests should help you finish successfully on the official TOEFL. For more on visualizing your goal, refer to the Answer Key on page 391.

Visualization Image for College Students

7 AUDIOSCRIPTS AND ANSWER KEYS

CHAPTER 2—READING STRATEGIES

PRACTICE 1

Five things that are wrong with the photo:

1. Headphones—put them on	"I am ready to listen. I am ready to speak."
2. Keyboard—move to side	"I am ready to write."
3. Paper—move to side	"I am ready to write."
4. Pencils—move to side	"I am ready to write."
5. Mouse—adjust	"I am ready to read."

PRACTICE 2

Three online suggestions:

1. *www.encyclopedia.org*
2. *www.infoplease.com*
3. *www.wikipedia.com*

Be careful! Some encyclopedias give you a free trial, but then you will be charged for membership online. Use the free encyclopedias, not the free trials.

PRACTICE 3

Pre-Columbian Civilizations

Although several other important cultures flourished, three major empires extended their influence over large regions in the Americas prior to the exploration and conquest by European powers in the sixteenth century. The Aztecs dominated the valley of Mexico where they intermarried with the Toltec nobility and systematically conquered smaller, weaker tribes from neighboring city states, consolidating their powerful kingdom under the rule of one chief. The subjugation of these regional city states allowed the Aztecs to exact tribute from more than half of the population of what is now Mexico, and assured their domination of Western Mesoamerica.

The Mayan culture stretched from El Salvador, Honduras, and Guatemala into central Mexico. During the classical period from 250 to 900 C.E., the construction of large urban areas identified their independent city states and dominated the landscape in Eastern Mesoamerica. Unlike the Aztec empire, there was no single Mayan political center. Because the Maya could not be overthrown by attacking a capital city where a centralized political system was in place, they remained strong and were able to survive invasions by competing tribes, exerting tremendous intellectual, if not occupational, influence throughout the region.

In what is now South America, the Incas established the largest empire of the Pre-Columbian cultures, uniting four regions in the Andes, including territories in what are now Ecuador, Colombia, Northern Chile and Argentina, Bolivia, and Peru. Ruling from a tribal base in Cuzco, the Inca emperor was considered the representative on Earth of one of the most revered gods, Inti, the sun god. The conquered tribes in the four regions maintained limited cultural identity under the rule of the Inca emperor but were, for the most part, integrated through peaceful negotiations and alliances with covenants of loyalty. Often the children of the ruling classes in the four regions were taken to Cuzco for their education, and daughters of the Inca nobility were married to powerful families in distant corners of the empire. If these peaceful methods proved inadequate, the formidable armies of the empire could easily subdue uncooperative tribes, execute the local leaders, and annex their cities.

PRACTICE 4

Stonehenge

Arguably, the most recognizable megalithic monument in Europe is Stonehenge on the Salisbury Plain in southern England. Like other henges, Stonehenge is a circular arrangement of large stones, surrounded by a ditch. Although it has not been determined exactly who was responsible for the construction of Stonehenge, it was probably built in stages over several hundred years, beginning about 3,000 B.C.E.

The last henge, completed about 1,500 B.C.E., was designed as a post and lintel structure in concentric circles. The outer ring, which is almost 100 feet in diameter, is constructed of huge sandstone or sarsen megaliths with smaller volcanic bluestones from Wales forming the inner ring. Inside the bluestones, a semicircle opens to a long avenue marked by uprights on an axis to the east identified by the Heel Stone, a large stone with a pointed top. A person standing at the center can view at the Heel Stone the exact spot where the sun rises at the summer solstice, the longest day of the year. It is assumed that Stonehenge must have been some kind of solar calendar or an early observatory.

Clearly, the most intriguing question involves the engineering methods that allowed early man to build such a remarkable structure. The gigantic stones in the outer ring are 22 feet high and weigh as much as 50 tons each. They have been traced to Marlborough Downs, 20 miles north of Stonehenge. Although most of the land is flat, at Redhorn Hill, the steepest part of the road, modern estimates suggest that more than 600 men would have been required to transport one of the stones uphill. The raising of each lintel, a stone beam that rests on top of two vertical stone posts, would also have required heroic effort. Based on the use of sledges, rafts, pulleys, and other machines available to Neolithic humans, work studies calculate that 30 million hours of labor would be the minimum to quarry the stone, transport it to the site, and complete the arrangement of the posts and lintels for the monument. The positioning of the stones to place the rising sun exactly over the Heel Stone during the solstice is also impressive.

PRACTICE 5

1. **D** The word *reliably* in the passage is closest in meaning to "dependably." "Hybrid varieties dependably produce full, sweet kernels . . . "
2. **A** The word *precise* in the passage is closest in meaning to "correct." ". . . the results . . . are not always correct."

Synonyms

analyze	study carefully, examine, investigate, evaluate
approach	method, technique
area	part, region, section, zone, subject
assess	measure, judge, consider
assume	suppose
authority	power, influence
available	obtainable, accessible
benefit	advantage, assistance
concept	idea, notion, thought, theory, model, hypothesis
consist	comprise, involve, entail
constitute	comprise, compose, form
context	background, framework, perspective, setting
contract	agreement, bond, treaty, deal
data	facts, information, statistics
define	describe, explain
derive	develop, originate
distribute	divide, dispense
economy	financial system, wealth, market
environment	location, nature, setting
establish	found, create, inaugurate, institute
estimate	approximate, guess
evident	easy to see, apparent
factor	consideration, aspect, feature, issue
finance	provide money, fund
formula	method, plan, procedure, principle
function	purpose, role
income	revenue, proceeds, earnings, profits, returns
indicate	designate, specify, signify, denote, show, demonstrate
individual	separate, discrete, distinct, specific, single
interpret	understand, clarify
involve	include, engage
issue	subject, topic, concern
labor	work, workforce, effort
legal	lawful, authorized, permissible, official
legislate	establish, decree, authorize, enact
major	most important, principal, chief, primary, foremost, main
method	approach, technique, process, system
occur	take place, happen, ensue, arise
percent	part, fraction, ratio
period	era, age, stage, epoch, phase, time
policy	course of action, rule, strategy, plan, procedure

principle	assumption, tenet, notion, theory
proceed	go on, continue, advance, progress
process	procedure, method, manner, means
require	need, compel, demand, expect
research	study, investigation, examination, inquiry
respond	answer, reply, react
role	position, function, responsibility, job
section	part, portion, unit, piece, segment
sector	area, segment, part, division, region
significant	important, major, substantial, vital, key, principal, critical
similar	like, alike, comparable
source	basis, cause, origin
specific	particular, exact, definite, certain, individual
structure	organization, construction, arrangement, composition, form
theory	hypothesis, speculation, assumption, premise, supposition
vary	differ, contrast, alter

PRACTICE 6

1. Although rare, the marriage of one woman to two men may occur when the two men involved are brothers.
2. Chemical digestion, which is carried on by certain protein-based molecules, reduces the size of the particles of food in the stomach.
3. Sumerian art before 4,000 B.C.E. centered around temples with towers and sculptures of their rulers.
4. Intangible property such as magical rituals or medicinal formulas are usually transferred before death because they require a period of learning.
5. It is mainly through the leaves of plants that the loss of water by evaporation occurs.
6. The response in the left part of the limbic system in the temporal lobes of the brain was greater when presented with a fearful stimulus as opposed to a pleasant stimulus.
7. It was during the Middle Ages that the science of compiling and editing a dictionary came into existence.
8. By far, the largest living group of plants with seeds that are unprotected by an ovary are pine trees.
9. Durkheim introduced the concept of a breakdown in social standards in the late 1800s to explain the response by society to the Industrial Revolution.
10. In spite of arguments to the contrary, altering the value of currency can cause a loss of stability due to an imbalance in supply and demand.

PRACTICE 7

The Missouri Compromise

The Missouri Territory was the first to be organized from lands acquired in the Louisiana Purchase. At the time that Missouri petitioned for statehood in 1819, Senate membership was evenly divided between states that were designated as slave-holders and those that were free. Clearly, the admission of a new state would destroy the balance, providing a voting majority for one side or the other. Although the Civil War was still forty years away, slavery was already a hotly contested topic in Congress with about equal numbers in favor and opposed to it. Since slavery had already been established in the Missouri Territory, it appeared that the scale would tip in favor of slave states.

During the long debate, the Territory of Massachusetts applied for admission as the state of Maine. Quick to recognize a way to settle the issue peacefully, Speaker of the House Henry Clay proposed a solution that has come to be called "The Missouri Compromise." Combining the two petitions for statehood, he fashioned a bill that allowed the Senate to retain equal numbers of Senators from slave and free states. Missouri would be admitted as a slave state at the same time that Maine was admitted as a free state. As part of the bill, Clay introduced a plan to maintain the balance of power between the two factions of the Senate as future territories petitioned it for statehood. A provision that divided the Louisiana Purchase at the southern boundary of Missouri, that is, 36 degrees and 30 minutes latitude, allowed for slave states south of the boundary and free states north of the border—with the exception of the Missouri Territory—to enter as a slave state because of its prior status.

Thomas Jefferson, now an aging statesman, saw the potential for a division in the union based on regionalism and the slavery question. By 1850, his prediction was very much a reality. Southern states threatened secession, and a second compromise was necessary to avoid it. Again brokered by Henry Clay with assistance from Stephen Douglas, it was a complicated effort consisting of five separate bills to deal with the land in the Southwestern Territories that were beginning to organize for statehood, all of which were south of the thirty-sixth parallel. The plan, commonly referred to as the Second Missouri Compromise, took four years to complete and was successful for less than a decade in forestalling the Civil War.

PRACTICE 8

Types of Play

Children engage in different types of play, and although it is helpful to categorize them, it should be understood that there is a great deal of overlap among them. Besides, several different methods of classification have been devised, a circumstance which adds to the confusion. However, it is useful to classify the major types that seem to appear in most of the lists.

The first type that is commonly identified is *sensorimotor play*, which begins in infancy and includes motor activities such as crawling, running, jumping, waving, or playing peek-a-boo. Sensorimotor play also includes opportunities for children to manipulate objects such as a rattle or a ball and allows them to enjoy physical stimulation as they explore the environment. These normal activities are easily observed as young children interact with people and objects at a very early stage of life.

➔ Another type of play that most psychologists have identified is *imaginative play*, which usually involves [make-believe situations.] Children pretend to be someone else or imagine an activity or place that is part of a [fantasy world.] Daydreaming is one form of imaginative play, but often children actually create involved scenarios and improvised dialogues. Since this happens after children are old enough to have memories, most adults recall playing "Make Believe."

Constructive play, also called *creative play,* is a third type. Examples include making music, either with real or toy instruments, creating art such as drawing, painting, molding with clay, building with blocks or other materials, as well as cooking and completing simple sewing projects. Of course, this can be structured in a classroom setting, or it can be a spontaneous event.

Finally, the fourth type of play, *cooperative play,* involves more structured activities which require interaction with other children in socially acceptable ways. One of the highest levels of cooperative play consists of games with rules, including not only sports like baseball, basketball, and soccer but also board games like checkers and Monopoly, or card games at varying levels of complexity.

➔ So what does this mean? The different types of play, in addition to teaching children social skills, also provide an [important way to build cognitive skills.] Children who are allowed to experience a wide range of play situations have demonstrated that they are more able to respond to unique or unfamiliar situations in real life. They react more acceptably in social settings and form more appropriate relationships.

According to Paragraph 3, which of the following activities is included in imaginative play?

- Ⓐ Playing peek-a-boo
- Ⓑ Running and jumping
- Ⓒ Building with blocks
- ● Making up fantasies

Paragraph 3 is marked with an arrow [➔]

According to paragraph 6, why is play important in childhood?

- Ⓐ It teaches rules for games.
- Ⓑ It allows children to explore the arts.
- ● It develops reasoning skills.
- Ⓓ It helps children choose friends.

Paragraph 6 is marked with an arrow [➔]

PRACTICE 9

What is different about the skeleton of a seahorse?
Word: skeleton

How do seahorses swim?
Word: swim

When was research data made available about the seahorse's head?
Numbers: date

Where are researchers studying the shape of the seahorse?
Capital letters: place

Who is the investigator using biomechanical analysis to understand the evolution of the seahorse?
Capital letters: name

How long does it take for seahorses to hatch?
Numbers: duration of time

How many young seahorses will survive to maturity?
Numbers: percentage

Seahorses

Seahorses are unique in many ways. Although they are bony fish, their skeletons are composed of a series of plates arranged in rings around which a thin skin is stretched, and they do not have scales. They swim upright, which is also a distinct characteristic, and they are often found at rest because they are very poor swimmers compared to other fish. Seahorses move through the water by rapidly fluttering a dorsal fin and steering with a pectoral fin positioned behind the eyes. While they are resting, they curl their prehensile tails around a stationary object like a monkey might do in a tree to stabilize itself.

One of the most striking characteristics of seahorses is their shape. Their tiny heads closely resemble the horse for which they are named. Even the spikey crown looks like a small mane. In 2011, researchers at the University of Antwerp in Belgium released data that included several interesting clues about the horse-like appearance. Using biomechanical analysis, Dr. Sam Van Wassenbergh and his team learned that the seahorse can capture its prey of small shrimp at a greater distance than can the pipefish, which, along with the seahorse, descended from a common ancestor. According to Dr. Van Wassenbergh, the arched head of the seahorse enables it to bend its head and extend it in a snapping technique that allows it to achieve a larger striking distance. The team believes that natural selection would have favored this foraging behavior, and encouraged the horse-like head to develop.

Reproduction in seahorses is also distinct from most other animals. It is, in fact, the male that carries the eggs in a special egg pouch that serves as an incubator until the small seahorses hatch fully formed. During gestation, which lasts from two to four weeks, the female visits the male for a few minutes every morning, then swims away. The number of young varies dramatically from as few as 5 to as many as 1,500, only .5 percent of which will actually survive to adulthood.

PRACTICE 10

According to the passage, all of the following stimuli were used in Pavlov's experiments EXCEPT

Ⓐ Footsteps

● A shock

Ⓒ A lab coat

Ⓓ A light

Notes

A ✔
B
C ✔
D ✔

According to the passage, classical conditioning has been useful in all of the following EXCEPT

Ⓐ Treatments for anxiety

Ⓑ Modern research

● Accidents

Ⓓ Advertising campaigns

Notes

A ✔
B ✔
C
D ✔

Classical Conditioning

Although the phenomenon had been observed prior to Pavlov's investigations, he was the first to appreciate the significance of *conditioned response*. An accidental discovery, conditioned response was observed while Pavlov was studying the gastric secretions of dogs in his laboratory. He noted that the sound of his associate's footsteps caused the dogs to salivate even before the food was brought in. Later, he was able to use other stimuli to evoke the same response. For example, by presenting a light followed by food, Pavlov could condition the dogs to salivate after the light was presented and before the food was offered. In this case, the salivation was an *unconditioned response* to food, the light was the *conditioned stimulus*, and the salivation was the *conditioned response* to the light.

Repeating the experiment in numerous variations, Pavlov determined that he could replicate the results with many neutral stimuli including a white lab coat, a bell, and a tuning fork as well as a light. By pairing the neutral stimuli with food, the dogs would salivate before the food appeared. Ultimately, he could elicit salivation by presenting only the neutral stimulus, without the presence of the food.

In these basic experiments, Pavlov established the basic terminology and methodology that continues to be used in modern classical conditioning experiments. The discovery that environmental events with no relationship to a given response could, through experience, cause the response through association, was a breakthrough in behavioral psychology.

Animals, like Pavlov's dogs, tend to associate stimuli with survival. Consequently, the use of food in animal experiments would be logical; however, human subjects may associate unrelated stimuli with complex emotions. People with unusual fears or phobias may have had an emotionally charged experience with which they have associated a harmless object. In that case, it would be appropriate to reverse the conditioned response. Pavlov called this reverse conditioning *extinction*, a process that is still very effective in treating anxiety.

Because conditioning is so basic to learned human behavior, commercial advertising has exploited Pavlov's research in order to influence purchasing behavior. Effective commercials often train the public to associate a product with a pleasant or desirable stimulus that evokes a positive response. In this case, the public has an unconditioned response to the product, the conditioned stimulus is a beautiful woman, a luxury lifestyle, or some other desirable image, and the purchase is a conditioned response.

PRACTICE 11

1. **Noam Chomsky proposed the language acquisition device (LAD) as an explanation for the learning of native languages by young children.**

   The theory assumes that children are born with an innate facility for acquiring any language.

      First language learning occurs too rapidly for such a complex system unless children have the LAD.

   Universal grammar, common to all languages, is present in the brains of human infants at birth.

      Word order for sentences in languages is restricted to a small number of possibilities.

   The LAD scans for input signals that allow children to focus and learn a specific language.

      Children do not simply repeat sentences that they hear from limited, often fragmented adult input.

2. **The Hubble Telescope has provided invaluable information that has virtually revolutionized the field of astronomy.**

   Hundreds of planets have been discovered in regions beyond our solar system.

      Two planets were found orbiting around a pulsar in a neighboring solar system.

   Scientists have found an unexplained force called dark energy, which is accelerating the expansion of the universe.

      The expansion of the universe seems to be speeding up instead of slowing down.

   The Hubble Deep Field reveals galaxies farther away than we have ever been able to see before.

      Galaxies vary in shape, color, and size, and their age goes back almost to the inception of the universe.

3. **Humans have been adorning themselves with jewelry from ancient times, using local materials.**

   Prehistoric peoples made jewelry of leather or grass, which was strung with shells, stones, or bones.

      Some of the jewelry has been preserved in caves and in buried sites of ancient cities.

   Ancient Western cultures preferred gold from as early as 3,000 B.C.E. for rings, earrings, and necklaces.

      Beautiful necklaces and earrings have been found in burial sites in Egypt, Greece, and Rome.

   In China, craftsmen used silver, which they enameled to create necklaces, bracelets, rings, and anklets.

      Jewelry designs by Chinese artisans included flowers, turtles, birds, and dragons.

4. **Bird nests built in trees are generally categorized on the basis of their style.**

   Some nests are constructed of mud mixed with the bird's saliva, shaped, and allowed to dry against the tree trunk or a large branch.

      Mud nests must dry out before each new layer is added to the nest or they would collapse.

   Birds take advantage of natural cavities in trees, enlarging and shaping them by chiseling or chewing with strong beaks.

      Woodpeckers use holes that have been excavated by insects, and improve them by pecking out additional spaces.

   A nest situated in the fork of a tree is typically built with grass, twigs, and other materials shaped into the shape of a cup.

      While some birds can complete a cup nest in one day, others require up to two weeks to construct their nests.

5. **The work of e.e. cummings does not conform to the usual conventions of poetry and language.**

   Cummings did not always capitalize the personal and place names as required by the rules for using capital letters.

      The author preferred to use lower case letters for his own initials and last name on his poems.

   Eccentric punctuation was characteristic of Cummings's unique literary style.

      The absence of punctuation was one of the ways that the poet departed from the traditional rules.

   Cummings's inventive formations of compound words make his poems both surprising and unique.

      Adjectives such as "mud-luscious" and "watersmooth" create images unlike that of any other poet.

PRACTICE 12

1. Drums were used in ancient Chinese culture for all ceremonial occasions . . . Weddings and funerals were among the most common ceremonies.

   Drums were probably used in Chinese weddings and funerals.

2. Fossils are the remains or impressions of plants or animals, embedded in rock . . . Petrified wood results when organic materials are replaced with minerals, thereby turning a tree into stone.

   Petrified wood is a fossil.

3. The great railroad hotels were built close to the railroad lines in the late nineteenth and early twentieth century at stops near landmarks of interest to tourists...The Empress Hotel in Victoria is one of the most elegant railroad hotels in Canada.

   Victoria must be a tourist site.

4. Unlike ferns, mosses and liverworts lack complex vascular tissue . . . The absence of an extensive vascular system restricts the size of plants.

   Ferns are probably larger than mosses and liverworts.

5. Studies at Princeton University suggest that it takes most subjects one-tenth of one second to form an impression of a stranger's face . . . Several minutes of exposure to the face does not change the initial impression, although subjects do report greater confidence about their judgments.

   In a job interview, the candidate will make a first impression in less than a second.

PRACTICE 13

1. There is a great deal of controversy about how to distinguish a lake from a pond, especially among the international scientific community.

   In other words, an internationally accepted definition of a pond cannot be agreed upon.

   This insert sentence restates the previous sentence.

2. Crater lakes are sometimes formed in extinct volcanic craters that fill up with fresh water more rapidly than they evaporate.

   Moreover, these fresh water lakes tend to have exceptional clarity because they lack inflowing stream water that would muddy them with sediment.

   This insert sentence includes the transition word "Moreover," which introduces additional information about fresh water lakes.

3. Near the end of the last Ice Age, glaciers began to retreat, leaving behind large deposits of ice in the depressions that they had gouged out, creating lakes as the ice melted.

   For example, in Minnesota, Lake Duluth was formed at the southern tip of Lake Superior, originally occupying a much larger area than Lake Superior does now.

   This insert sentence provides an example of a lake that was created.

4. A glacial lake is a lake that is created by a melted glacier, as for example, a kettle lake.

   In contrast, an oxbow lake is formed when a river meanders and becomes very curved.

   This insert sentence contrasts an oxbow lake with a glacial lake.

5. Lakes formed by the impact of a meteor have distinct characteristics. First, they tend to be perfectly round. In addition, a raised rim often outlines the depression.

   Finally, the water is very pure, since it is fed by rain or snow.

   This insert sentence should be at the end of a series of ideas.

PRACTICE 14

1. After the fall of the Mycenaean civilization, villages began to unite in order to form strong trading centers, which eventually evolved into the Greek city-states.

   The Mycenaean civilization fell.
   Villages began to unite in order to form strong trading centers.
   Greek city-states evolved.

2. Although the peer group in school is significant in the social experience of a growing child, the family unit remains fundamental throughout the entire life span of an individual.

   The peer group in schools is significant in the social experience of a growing child.
   The family unit remains fundamental throughout the entire life span of an individual.

3. Bones are not considered a solid structure because they generally have a large cavity in the center to accommodate marrow.

   Bones are not considered a solid structure.
   They generally have a large cavity in the center to accommodate marrow.

4. Changes in technology have influenced the advertising industry dramatically in the past few decades, as for example, the pop-up ad on the Internet.

   Changes in technology have influenced the advertising industry dramatically in the past few decades.
   The pop-up ad on the Internet is an example of the influence.

5. Studies with human subjects have confirmed that eating 30 percent fewer calories can increase one's life span as long as the essential nutrients are included in the lower-calorie diet.

   Studies with human subjects have confirmed that eating 30 percent fewer calories can increase one's life span.
   The essential nutrients are included in the lower-calorie diet.

6. The first printing presses were inspired by screw presses that were used in agricultural industries for pressing grapes for wine and olive oil seeds for oil.

   The first printing presses were inspired by screw presses.
   Screw presses were used in agricultural industries for pressing grapes for wine and olive oil seeds for oil.

7. To survive and thrive, coral reefs require water that is at least 20 degrees Celsius and adequate sunlight as well as clear water, less than 100 meters deep.

   To survive and thrive, coral reefs require water that is at least 20 degrees Celsius.
   Coral reefs require adequate sunlight.
   Coral reefs require clear water, less than 100 meters deep.

8. Art historians suggest that some important works that were previously attributed to male artists may actually have been created by women because they were unable to market their art under their own names.

   Art historians suggest that some important works were previously attributed to male artists.
   These important works may actually have been created by women.
   Women were unable to market their art under their own names.

9. Many successful authors have had their books rejected by publishers, including J.K. Rowling, the creator of the Harry Potter series, whose original manuscript was refused by five publishers.

   Many successful authors have had their books rejected by publishers, including J.K. Rowling.
   J.K. Rowling was the creator of the Harry Potter series.
   The original manuscript of the Harry Potter series was refused by five publishers.

10. Although they are often prized for their beautiful coloring and their lovely songs, songbirds are not only an aesthetic addition to the environment but also a practical means to control insects and weeds without chemical pesticides.

    Songbirds are often prized for their beautiful coloring and their lovely songs.
    Songbirds are an aesthetic addition to the environment.
    Songbirds are a practical means to control insects and weeds without chemical pesticides.

PRACTICE 15

QUESTION

What can you do to stop saying or hearing the words in your head?

- ✔ Put a pencil in my mouth
- ✔ Put my finger over my lips

League of Nations

The League of Nations was established at the end of World War One in a covenant agreed to in the Treaty of Versailles. After a long and devastating global war, the League was formed in an effort to bring stability to the world. The mission of the new organization was to promote international cooperation and to maintain world peace. The League was based in Geneva, Switzerland, because Switzerland had been a neutral nation during the war.

In the event of an international incident, it was agreed that member nations would not go to war before submitting their dispute to the League, which would assist the nations involved to discuss their differences in an orderly forum. In addition, the League could impose a warning to an aggressive nation with consequences in the form of sanctions if that nation did not cooperate. Economic sanctions in the most severe form would include a trade embargo by League members. When economic sanctions were imposed, theoretically, it would cause distress within the offending nation to such an extent that the population would be motivated to pressure the government to accept the League's terms before the economy was bankrupted. If economic sanctions failed, then the League could impose order through military force.

The obvious flaw in the organization was the fact that the League did not have an army and the only countries in the League with a military presence were Great Britain and France, both of which were weakened by the recent war effort. Neither Germany nor Russia was allowed to participate, and the United States chose not to join. Therefore, although the League had lofty ideals—to preserve world peace—there was little chance for it to succeed on a political level because it did not have a powerful enough presence. Nevertheless, the League did have some victories in the social arena. The Health Organization made progress in providing fresh water wells and eliminating leprosy as well as improving the status of women and attacking the issue of child labor worldwide. When the League is evaluated and criticized, it is important to remember that it was the League that began many of the social programs continued through the United Nations today.

PRACTICE 16

Question

What will help you find larger chunks of text to focus on with your eyes?

✔ Mark the meaningful phrases in practice readings.
✔ Try to focus on an entire line in a column of text on practice tests.

Bipedalism

Bipedalism, that is, walking upright on two legs, is one of the characteristics of human behavior that distinguishes us from our closest relatives on the evolutionary chart—the chimpanzees, gorillas, and other nonhuman primates. Fossil evidence suggests that bipedalism occurred soon after the evolutionary divergence of human beings from apes, and is considered an important development, one that defined us as *Homo sapiens*.

Many theories have been put forward to explain bipedalism, but several of them are fairly far-fetched, and only about half of them have gained acceptance among the scientific community. One of the more popular theories assumes that walking on two legs made it possible for early humans to use their arms to reach for food and carry food. This theory argues that chimpanzees occasionally walk upright when they are grabbing food from overhead branches or carrying food for later consumption. Thus, bipedalism actually supported survival because it made it possible to gather food more efficiently or to carry food to others in the group.

Another theory, again citing evolutionary bipedal development as a survival mechanism, asserts that raising the head to a higher level while walking upright extended the field of vision, and that made it possible for human beings to see both predators and prey at a greater distance. Therefore, the humans that stood and walked upright would have had an advantage and natural selection may have favored them.

Yet another theory suggests that raising the body above the ground level helps in the dissipation of heat because more favorable breezes adjust the body temperature. Again the evidence for this theory relies on observation of nonhuman primates who walk upright in order to cool down.

Finally, there is the theory that stone tools and weapons were easier to use and more efficiently carried when human beings had freed their hands for that purpose. And this particular theory has a lot of support among anthropologists who associate the use of tools with the evolutionary leap from ape to human.

PRACTICE 17

QUESTION

How can you avoid rereading when you are looking at a page?

✔ Cover the previous lines with a paper.

How can you avoid rereading when you are looking at a screen?

✔ Scroll down to the current line to hide the previous lines.

Be sure to practice reading text on a computer screen to practice this skill.

Musicology

Music history, also referred to as historical musicology is the study of composition, performance, and critique over time. For example, a historical study of music would include biographical information about each composer and the way that a specific piece relates to other works, the development of styles of music, the relationship of music to social life, and the techniques associated with performance.

In theory, music history could treat the study of any type of music, but in practice, it has heavily favored Western music, and in particular, classical music. This perspective is problematic for two reasons. First of all, it dismisses music from other cultures. Because classical music was developed in nineteenth century Europe, it is representative of Western culture. But Eastern music has a rich, but very different tradition, which is virtually absent from the field of music history. Another problem is that music history arbitrarily ignores popular music, even Western music that has been composed post nineteenth century. Furthermore, the vocabulary and therefore the way that we talk about music is restricted because classical music is different from many other types of music in different cultures and time periods.

Consequently, when we look outside of Western classical music, we tend to look through the prism of classical music and often judge other music as inferior because it does not match the model that has been set up. For instance, classical music has a long tradition of harmony, but rhythm is not a key feature. In contrast, popular music, which has a very simple melody but a complex rhythmic structure, may be viewed as less valuable because it doesn't fit the mold. Another issue is the absence of notation in many non-Western forms of music. The fact that the rise of classical music coincided with the practice of musical notation elevates music that has a written tradition. Jazz, for example, would be considered less important because it relies on improvisation instead of sheet music.

PRACTICE 18

At the end of 10 minutes, you should have completed the reading passage and the five questions with at least four correct answers.

1. C
2. B
3. D
4. B
5. B

PRACTICE 19

Were you able to eliminate some of the answers? Which answer did you guess?
Remember that the first guess is usually the best choice.
The correct answer for this practice question is C.

PRACTICE 20

1. What happens when you click on the **Review** button?
   It takes you to the Review screen.

2. What is the purpose of the **Review** screen?
   It shows you whether you have answered every question.

3. When should you click on the **Review** button?
   At the end of each passage.

4. How often should you use the **Review** button?
   You should try not to use it very often because it takes time.

5. Why would you leave a question blank on the Reading section?
   By accident or because you want to return to it later.

CHAPTER 3–LISTENING STRATEGIES

CD 1, Track 1

Welcome to *Outsmart the TOEFL: Test Strategies and Tips*. Many books help you prepare for the TOEFL, but this book, and the audio that accompanies it, is different from all the others. *Outsmart the TOEFL: Test Strategies and Tips* helps you prepare for the TOEFL before you take the test and it helps you outsmart the TOEFL while you are actually taking the test. How is this possible? By using this book, you will learn the strategies that you need. Let's begin!

PRACTICE 1

CD 1, Track 2

Now listen to part of a lecture in a biology class. The professor is talking about migration. [You will also hear student responses to tasks on the Speaking section.]

Weighing in at only 3.5 ounces, the tiny Arctic Tern has the longest migration of any bird. Hard to believe, but the Greenland Institute of Natural Resources has documented a trip of nearly 45,000 miles from the northern coast of Greenland to the shores of Antarctica and back again. Interestingly, although we have known for a long time that the tiny Tern was engaged in a very long migration, we were not able to use the same types of instruments that we have routinely attached to larger birds like geese and penguins because the size and weight of the Tern made it impossible for them to carry the equipment on their flights. But with the development of smaller and smaller tracking devices, some about the size and weight of a paper clip, now scientists have been able to attach the instruments to the legs of the Terns and collect data that proves the length of their incredible journey.

Okay then, almost all research returns some surprises, and the Greenland Institute study was no exception. First of all, the scientists were surprised that the Terns do not fly by the shortest route for their migration, in fact, choosing to meander over the Atlantic Ocean in a curving route that adds almost 1,000 miles to the route but takes advantage of the wind currents that allow them to glide for relatively long periods of time. Another interesting point is the fact that the Terns use these wind currents to gain lift on the return trip, allowing them to make the migration north in about half the time that they need for the migration south. And the fact that the Terns stop for a month in the middle of the journey to feed on krill in the North Atlantic was not anticipated. After the stop, then half of the birds follow the coast of Africa and the other half fly along the coast of South America, meeting in Antarctica at the end of their migrations, where they live until it is time to go back to Greenland to breed and nest . . .

Example Notes

Arctic Tern—3.5 oz
Longest migration
45,000 mi Greenland—Antarctica—Greenland
smaller devices—size paper clip—possible to track
Not shortest route—1000 extra mi/ glide wind currents
N ½ time as S
Stop 1 mo mid-Atlantic—feed on kill
½ Africa/ ½ S America—meet Greenland/breed, nest

Speaking Response

My favorite way to spend a free day is to relax at home, go out with my friends, and talk with my family. First, I'd sleep late because that isn't something I can do on school days. I might make a big breakfast, which would be an indulgence since I usually have to run out the door with just an energy bar to eat. I'd have two cups of coffee while I watched the news on my laptop and then I'd call my friends to make plans for later in the day. We could go shopping or just hang out. That evening, I'd meet my friends for dinner at a campus restaurant and after that, we'd all go back to my apartment to watch a movie. When everyone had gone, I might talk to my family on skype. It would be early morning in my country, and it would be the end of a perfect day here.

Speaking Response

According to the lecturer, Art Deco was a rather eclectic style that originated in Paris in the 1920s. The style influenced many areas of artistic expression, including painting, fashion and interior design, and architecture. Art Deco is interesting because it integrated elements from many different countries. Greece, Egypt, Mexico, and Africa contributed to the style, probably because, at the time, the interest in archeology was intense. Geometrical shapes were also a recurring theme in the movement. Some of the materials that were used in Art Deco were aluminum, chrome, and stainless steel. Although many fine examples of Art Deco architecture are found throughout the world, downtown Los Angeles, along Wilshire Boulevard has a large concentration of Art Deco Buildings.

PRACTICE 2

CD 1, Track 3

Now listen to part of a lecture in an art history class. The professor is talking about portrait painting.

From the colonial era well into the nineteenth century, portrait painting was a very popular art form in the Americas. Many early painters were not formally trained. In fact, a lot of them were sign painters or house painters who had taught themselves to paint likenesses which they sold to a middle class that was beginning to have more discretionary income. But by, oh, after the Revolutionary War, a number of portrait painters had separated themselves from the more mundane tradesmen and, uh, had elevated their status to artist. Some had studied in Europe where they visited the grand museums and became familiar with the works of the great masters. And Gilbert Stuart was, in my opinion, first among them. He had studied with Benjamin West in Europe and then returned to colonial America to work.

Now remember that this was before the invention of the camera, and portraits served to document the time period. So Gilbert Stuart painted many portraits that depicted historical achievements and life after independence. Among them were three portraits of George Washington, which he painted from life, and then using the originals as his models, he produced more than one hundred replicas of them. One of the portraits, commissioned by Martha Washington as part of a pair of portraits of her and her husband for their home would become probably Stuart's most famous portrait. It's called the Athenaeum Portrait, and I'm sure that you have all seen it because it was chosen for the engraving on the U.S. one-dollar bill. Interestingly enough, the original Athenaeum Portrait was left unfinished so that Stuart could continue to use it as a model for the many copies that he sold, promising Mrs. Washington that the original would be hers when he had completed it. Of course, he probably had no intention of ever completing it, and it hung in its unfinished state in the Boston Athenaeum for more than 150 years.

Actually, painting Washington's portrait was a shrewd move because images of Washington were in high demand not only in the Americas but also in Europe and Stuart used the images to gain recognition and a comfortable income. Stuart also painted portraits of the next five presidents and three first ladies . . .

Example Notes

Colonial—19th C	PP no training—sign/house p middle class
Rev War	artists studied Europe Gilbert Stuart w/ Benjamin West
Before camera	PP document period/historical

3 PP George Washington/life—100 replicas
Pair commissioned by M Washington—Athenaeum—engraving $1 original unfinished—Boston Athenaeum 150 yrs

Smart—images W in demand America/Europe
Stuart gained recognition—PP next 5 pres/3 first ladies

PRACTICE 3

CD 1, Track 4

Now listen to part of a lecture in a zoology class. The professor is talking about owls.

Owls are [illegible] in many ways, but it is probably their eyes that have attracted the most attention. Unlike other birds, owls have eyes in the front of their heads, like people. Also like humans, owls have [illegible] vision, which allows them to see the same object with both eyes. This is an important advantage for catching prey because it contributes positively to depth perception and [illegible] distances. The size of the owl's eyes is also unusual among birds. Although a great horned owl may be two feet tall, the eyes are as large as a six-foot-tall human and may account for as much as 5 percent of the animal's body weight.

The [illegible] shape of the eyes in owls is superior in many ways to the round ball found in most other animals. The fact that their eyes are set permanently in the [illegible] prevents the tubes from moving, which is why owls must turn their heads to change the angle of their perspective. The owl's neck, with twice the number of vertebrae as a human's neck, more than [illegible] for the [illegible] eye socket. Using their long and very flexible neck, the owl can turn its head around 270 degrees from its forward-facing position. Although this is not quite a full turn, it does allow the owl to look behind without [illegible] the rest of its body.

Owls have extraordinary night vision because they can [illegible] the pupils of their eyes almost to the width of the entire eye to let the maximum amount of light in. It was once thought that they were blind at night; however, scientists have proven that owls have excellent vision during the day as well. Some species of owls can see better than humans do in bright light.

In general, owls have black, brown or yellow eyes. To protect their eyes, they have three eyelids. The upper lid closes when the owl blinks, the lower lid closes when the owl is sleeping, and the third lid closes [illegible] across the eye from the inside to the outside. This thin lid cleans and protects the surface of the eye . . .

Example Notes
Some of the words in this lecture have been beeped. You should still be able to understand the main ideas without recognizing every word.

Topic: Owl's eyes

Major Point 1: Owls have eyes in front of heads
Sees w/both eyes
Depth + distance
Advantage/prey
Large size

Major Point 2: Tube shape
Superior to round
Permanent—don't move
Flexible neck 270 degrees

Major Point 3: Night vision
Adjust pupils
Maximum light in

PRACTICE 4

CD 1, Track 5
Now listen to part of a lecture in a physics class. The professor is talking about waves.

Maxwell's original discovery that light is an electromagnetic wave produced many questions. Uh, they began to arise because when Maxwell peered into the universe, visible light was the only example of electromagnetic waves, but . . . but visible light has an extremely narrow range of wavelengths. Let's see, uh, about 390 to 710 nanometers. But according to his equations, well, electromagnetic waves could exist at any wavelength and therefore, at any frequency. So you see the problem. Maxwell began . . . to . . . to wonder . . . what had happened to the rest of the waves. I mean, there should be more than uh, the equations would have predicted more kinds of electromagnetic waves than just light waves. And, uh, these waves, these unseen waves, would move at the speed of light and would have, uh, would be exactly the same as visible light, except . . . except for the differences in wavelength and frequency.

So, uh, so about twenty years later, Hertz began researching waves that we now call radio waves, and since then, well all kinds and sizes of waves have been discovered, uh, just to mention a few, microwaves, infrared, visible radiation, ultraviolet rays, X-rays, and gamma rays. Okay then, all of these waves, despite the frequency or wavelength, all of these waves form the electromagnetic spectrum, with radio waves at the low frequency end of the spectrum to gamma radiation at the short wave end of the spectrum. The micron is the basic unit for measuring the wavelength of electromagnetic waves, so going back to the longest waves, the radio waves, these can have wavelengths of many kilometers, and the gamma waves, the shortest waves, they have wavelengths of 6 microns, or even less . . .

Example Notes

Maxwell Light = elec mag wave

Visible L narrow wavelengths
390–710 nano

M equations predict more

Hertz 20 yrs radio waves
Later micro/ infrared/ radiation/ ultraviolet/ X-rays/ gamma = electromagnetic spectrum

micron unit
radio ➔ gamma
long kilometers short 6 m

PRACTICE 5

CD 1, Track 6

Listen to part of a lecture in an art class.

Before we begin, some of you have come in to see me about the presentations so let me take a few minutes to clarify a few things that seem to keep cropping up. First, I don't expect the presentations to be longer than 10 minutes. And if you have one person do the actual speaking, that's okay as long as everyone has made a contribution to the project. So someone may choose to do the handout and someone else may contribute by researching the topic or by making visuals for the speaker to use while presenting. Hope that helps, but if not, you can find me in my office during office hours. I'm happy to help you individually or as a group.

Good. Okay then. With the expansion of the World-Wide Web, a whole industry has developed around an area of graphic design that was unheard of until very recently, that is, web-site design. In this sense, the computer has not only supported designers in their traditional fields but has also created opportunities for designers to work in a completely new space—the computer itself. Design for the web relies on such traditional design models as advertising, magazine layout, and incorporates the potential for animation and interaction.

Perhaps it is John Maeda at the Massachusetts Institute of Technology who has done the most to merge the field of computer design with that of engineering. He rejects the tendency of graphic artists to use design software that can be purchased commercially, pointing out that they are limiting their artistic creation to the specifications offered them by the software designer. Maeda encourages graphic designers to learn how to program the computer themselves. To help them, he has published a groundbreaking work, *Design by the Numbers*, a book that demystifies the technology behind computer art, and shows designers how to do it. But even though he is working with the most modern technologies, Maeda often advocates a conservative approach to the visual art that is produced. His work promotes elegant visuals and communicative text, the principles that have formed the foundation of graphic design for centuries . . .

Example Notes
Signal words: Before we begin
Topic for classroom business: Presentation

Transition words: Good. Okay then.
Topic for this lecture: Web design

PRACTICE 6

CD 1, Track 7

Listen to part of a lecture in a sociology class.

Yesterday we concluded our discussion of preindustrial societies. As you will recall, a preindustrial society has a . . . a . . . social structure that is supported primarily by an agricultural economy and the production of limited goods by craftsmen. We talked about the fact that a rigid social class system limited the interaction among groups to specific divisions of labor. And . . . and that underlying all of these characteristics—at the heart of preindustrial societies really—was a loosely organized group of parochial communities. Transportation and communication systems were simple and as we said, information about the world was restricted, that is, little dissemination of knowledge or culture outside of the villages occurred. So the family unit and membership in a church or a village defined the person.

With that in mind, let's talk about *modernization,* a term that sociologists use to refer to the process of social and cultural change that is initiated by industrialization and results in a division of labor with accompanying differentiation in social groups. Changes as a result of modernization may be viewed as both positive and negative. Advances in communication and transportation, medical care, and a higher standard of living as measured by income and material comfort are cited as, well, the rewards of modernization. Pollution, stress, and conflict among social groups are among the . . . negative consequences.

Whether we support it or not, modernization is typical of most societies, and is characterized by three trends. The most obvious is the decline or even disappearance of small, traditional communities. Industrialization tends to diminish the importance of, of . . . the family unit . . . in favor of secondary groups such as colleagues at work. Also important is the multiplication of bureaucracies. When that happens, informal interactions among neighbors are replaced by impersonal communications. There is also a general tendency for traditional religious institutions to lose their central role in society. With bureaucratization and a loss of religious direction, many people feel that they are losing control of their own lives. Some will react by forming new communities and groups to replace the family and religious structure that have crumbled . . .

Example Notes
Signal words: Yesterday we concluded our discussion
As you will recall
We talked about
Topic for previous lecture: Preindustrial societies

Transition words: With that in mind, let's talk about
Topic for this lecture: Modernization

PRACTICE 7

CD 1, Track 8

1. Professor: Hey Linda. What's up?
   Student: If you have time, I'd like to hear a little more about the Semester Abroad—the program that you mentioned in class yesterday.

2. Student: Excuse me, Professor Philips.
   Professor: Oh, hi, John. Come on in. What can I do for you?
   Student: It's about my project. I was wondering whether you could take a look at it, uh, at what I have so far.

3. Professor: Hi, Elizabeth. How are you doing?
   Student: Fine. How are you?
   Professor: Doing great, thanks. How can I help you?
   Student: Well, if it's not too late, uh, I'd like to sign up for the field trip. I thought I had to work, but I traded days off with one of the other librarians so now I can go!

4. Student: Excuse me.
   Professor: Oh, hi, Steve.
   Student: Could I talk with you?
   Professor: Sure. What seems to be the problem?
   Student: I think I owe you an explanation, uh, for yesterday, for being late to class. I tried to talk with you after, but uh, there were a lot of other students around. I . . . I overslept.

5. Professor: Hey Kathy. What did you want to talk about?
   Student: Well, I was hoping that you might, uh, that you would let me use your name as a reference, for my application to graduate school.

Example Notes

1. ***Greeting:*** Hey Linda. What's up?
   Purpose: Info semester abroad

2. ***Greeting:*** Excuse me, Professor Philips.
   Oh, hi, John. Come on in. What can I do for you?
   Purpose: Look at project

3. ***Greeting:*** Hi, Elizabeth. How are you doing?
   Fine. How are you?
   Doing great, thanks. How can I help you?
   Purpose: Sign up field trip

4. ***Greeting:*** Excuse me.
   Oh, hi, Steve.
   Could I talk with you?
   Sure. What seems to be the problem?
   Purpose: Explanation late/overslept

5. ***Greeting:*** Hey Kathy. What did you want to talk about?
Purpose: Name as reference

PRACTICE 8

CD 1, Track 9

Listen to part of a lecture in a literature class. The professor is talking about the history of motion pictures.

Melodrama is a sentimental play that includes both music and drama. The characters featured are, well, stock characters really, stereotypical characters like the gallant hero, the heroine who must be rescued, and the cruel villain. And there is almost always a minor character to provide some comic relief. Now everyone tends to be either all good or all bad. The evil villain is the central figure, and the other characters respond to his diabolical plans. So the plot revolves around the villain. And the plot is usually exaggerated and sensational with spectacular scenes and sets and a lot of action. The outcome is predictable. After a long series of trials and reversals of fortune, the heroine is saved and good triumphs over evil. For the most part melodrama disappeared as a theatrical form in the early 1900s, but it did survive in silent films accompanied by a highly emotional musical score. The *Perils of Pauline* is the clearest example of melodrama transferred to film. It is a serialized melodrama shown in about 20 episodes of 30 minutes each in which a different villain menaces the beautiful heroine, but, of course, at the end of each episode, Pauline is rescued or somehow escapes her fate. Poetic justice is at the heart of melodrama, which conforms to the moral code of the period. After all the suspense and emotional twists and turns, good must be rewarded and evil must be punished . . .

Example Notes
Blackboard definition: Melodrama sentimental play—music/drama

Explanation: stereotypical characters—hero, heroine, villain
Plot—sensational/predictable good over evil
Disappeared early 1900s Continued silent films *Perils of Pauline*

PRACTICE 9

CD 1, Track 10

You are not expected to know which answers are correct. Just choose the number of answers to select.

1. Which of the following marine animals are Cetaceans? 2 answers
   - ☒ Dolphins
   - ☐ Sharks
   - ☒ Whales
   - ☐ Seals

2. According to the professor, which of the following days of the week were named for the Norse gods? 2 answers
   - ☒ Tuesday
   - ☒ Thursday
   - ☐ Saturday
   - ☐ Sunday

3. According to the passage, what do we know about oil paint? 1 answer
   - ○ It does not allow the artist to blend and refine colors.
   - ○ It tends to fade to a lighter color with age.
   - ● It is possible to paint over mistakes on the canvas.
   - ○ It dries evenly and quickly after application.

4. Which of the following are characteristic of viruses? 2 answers
   - ☒ They are much smaller than bacteria.
   - ☒ They must grow within a host.
   - ☐ They have a nucleus in their cell structure.
   - ☐ They respond to antibiotics.

5. According to the passage, which Canadian provinces are included in the Maritimes? 1 answer
   - ○ Newfoundland
   - ○ Quebec
   - ● Nova Scotia
   - ○ Yukon

PRACTICE 10

CD 1, Track 11

Listen to part of a lecture in a biology class. The professor is talking about cloning.

Cloning is one of the most controversial issues being discussed in the scientific community. As you know from your textbook, a clone is genetically identical to the cell donor. And every cell in an organism contains the complete DNA blueprint for producing an adult. But let me clarify something: During the development of a fertilized egg, most of the genes are switched off while the cells differentiate. So the point is that cloning actually *reestablishes* the DNA process. In the case of Dolly the sheep, for example, an unfertilized egg was taken from an adult Scottish Blackface sheep and the DNA was removed from the nucleus. Then a cell from another adult sheep, a Finn Dorset, was fused with the egg, using an electrical shock. The cell was taken from the mammary gland, I believe. Anyway, the egg then began to divide, which is the normal process of replication, and the embryo that resulted was implanted into the embryo of another adult sheep to develop to maturity. What is important here is that even though scientists didn't yet know how to switch the genes on and off in a cell, the unfertilized egg was able to make that happen.

So the result was Dolly, a sheep that was identical to the animal donor of the cell but different from both the animal that donated the egg and the animal that carried the embryo. And the experiment proved that a cell taken from a specific part of the body could, in fact, reproduce a complete organism, in this case, a sheep. But remember, the reprogramming process is not perfect and embryos produced by nuclear transfer, that is removing the cell nucleus from an adult cell and transferring it into an unfertilized egg that has had the nucleus removed, well, this is not a perfect process. Still, in Spain a wild mountain goat, a Pyrenean Ibex, which was on the official list of extinct animals, the Ibex was cloned from skin samples that had been kept in liquid nitrogen. Although the newborn Ibex died due to defective lung function, the accomplishment is tremendously important because it suggests that cloning may be one answer to reestablishing species that have become extinct or endangered. And that could be . . .

Example Notes

1. Signal words: But let me clarify something:
   Important information: During the development of a fertilized egg, most of the genes are switched off while the cells differentiate.

2. Signal words: So the point is that
   Important information: cloning actually *reestablishes* the DNA process

3. Signal words: What is important here is that
   Important information: even though scientists didn't yet know how to switch the genes on and off in a cell, the unfertilized egg was able to make that happen

4. Signal words: But remember
   Important information: the reprogramming process is not perfect

5. Signal words: the accomplishment is tremendously important
   Important information: because it suggests that cloning may be one answer to reestablishing species that have become extinct or endangered

PRACTICE 11

CD 1, Track 12

Listen to part of a lecture in a child development class. The professor is talking about infancy.

So babies are born with certain reflexes that exercise control over a newborn's movement. And reflexes are genetic . . . and . . . and . . . necessary to survival. Okay then. Reflexes are reactions to stimulation that are automatic. In other words, they are beyond the infant's control. But motor skills are developmental. Let me repeat that: Motor skills are developmental. Gross motor skills require large muscle activity, for example, lifting the head, rolling over, and sitting up and eventually, crawling and walking. So the development of these gross motor skills, they usually follow a sequence. And because the larger muscles develop before the smaller muscles, well, gross motor development forms the basis for developing fine motor skills.

Now, fine motor skills, those include the ability to grasp objects and manipulate them as well as finer eye-hand coordination. This begins with something called the *palmer grasp*. That's when an infant grips something with the whole hand. And they do that for about the first year. Then they start to grasp smaller objects with their thumb and forefinger. That is called the *pincer grip.* Now at this point in their development, an infant is moving the wrist to reach, then rotating the hand and coordinating the thumb and forefinger, which really requires a lot of fine motor skills.

One very interesting aspect of motor skills development is the fact that it generally proceeds from top to bottom, and by that I mean that infants learn to control their head, first, then trunk, hands, and legs. Infants progress in their motor skill development because, of course, their muscles are developing, but in order to develop gross motor skills and fine motor skills, the brain must also mature and develop. And it seems that infants are all hardwired with the developmental sequences in place. To say that another way: They are born with the sequences, but they need the brain and muscles to develop in order to activate them. Let me say that again. They are born with the sequences, but they need the brain and muscles to develop in order to activate them. Then all that they need is the opportunity to practice using their motor skills as they interact with the environment . . .

Example Notes

1. Signal words: In other words
   Restatement: reflexes beyond infant's control

2. Signal words: Let me repeat that
   Repetition: motor skills are developmental/sequence
   palmer grasp/pincer grip

3. Signal words: And by that I mean
   Restatement: top to bottom—head, trunk, hands, legs

4. Signal words: To say that another way
   Restatement: born w/ sequences—brain, muscles develop/activate

5. Signal words: Let me say that again
   Repetition: born w/ sequences—brain, muscles develop/activate

PRACTICE 12

CD 1, Track 13

Listen to part of a lecture in an anatomy class. The professor is talking about the way that the eyes adapt to changes in light.

When we enter a dark room, it takes our eyes about, oh, 30 to 40 minutes to achieve a complete adaptation. So how does that happen? Well, it actually takes place as a . . . a chemical reaction in the rods and cones. As you will remember from the textbook, the rods and cones are both photoreceptor cells, but the cones are not as sensitive as the rods, so they . . . the cones, I mean . . . they tend to adapt first. And it takes about 10 minutes for them to fully adapt. But the rods will continue adapting for as long as half an hour longer . . . up to 40 minutes from the time that we go into the dark room.

Okay then. The rods can't perceive color, so it makes sense that you can't discriminate color easily in dim light. But the cones are sensitive to light, especially yellow light. What does this mean? In the real world? Well, first let's get some terminology down. The primary use of the cones is called a photopic system, whereas the rods, the primary use of the rods, is called a scotopic system. So as we transition between light and dark, our eyes are shifting between their reliance on the rods and cones. To put that in scientific terms, we are transitioning between photopic and scotopic systems.

But what if it's important to keep both systems active? Let's say, in research settings with animals. Since many laboratory animals, like, uh, rats and mice, they have fewer cone photoreceptors, so they have limited photopic vision. But they are nocturnal animals so they are more active in the dark. But the human researchers need to observe the animals and perform procedures, or . . . or . . . read the settings on equipment. So here's the solution: Red lights allow the animals to perceive their environment as dark, while the human researchers, who have a cone that is sensitive to long wavelengths . . . they can see well enough to read their instruments or perform the tasks that they need to do in the laboratory setting. That's why you see laboratories with red lights or researchers using red glasses . . .

Example Notes

1. Rhetorical question: So how does that happen?
   Answer: Well, it actually takes place as a . . . a chemical reaction in the rods and cones.

2. Rhetorical question: But the cones are sensitive to light, especially yellow light. What does this mean? In the real world?
   Answer: So as we transition between light and dark, our eyes are shifting between their reliance on the rods and cones.

3. Rhetorical question: But what if it's important to keep both systems active?
   Answer: So here's the solution: Red lights allow the animals to perceive their environment as dark, while the human researchers, who have a cone that is sensitive to long wavelengths . . . they can see well enough to read their instruments or perform the tasks . . .

EXAMPLE SCRIPT 13

CD 1, Track 14

Neutral: That's a good deal for a meal plan.
Surprised: That's a good deal for a meal plan.

Neutral: That's not a problem.
Surprised: That's not a problem.

Neutral: My advisor has to sign this form.
Surprised: My advisor has to sign this form.

Neutral: The door to the classroom is locked.
Surprised: The door to the classroom is locked.

Neutral: The room has been changed.
Surprised: The room has been changed.

Neutral: Tuition has gone up again.
Surprised: Tuition has gone up again.

Neutral: My name isn't spelled correctly on the transcript.
Surprised: My name isn't spelled correctly on the transcript.

Neutral: He let you borrow his notes.
Surprised: He let you borrow his notes.

Neutral: The time limit's up.
Surprised: The time limit's up.

Neutral: The test is today.
Surprised: The test is today.

PRACTICE 13

CD 1, Track 14 Resumed

1. He's a student. ☐ Neutral ☑ Surprised
2. Susan is working full time at the library. ☑ Neutral ☐ Surprised
3. The bookstore is open. ☑ Neutral ☐ Surprised
4. I owe a late fee. ☐ Neutral ☑ Surprised
5. My advisor didn't approve it. ☐ Neutral ☑ Surprised
6. The class is off campus. ☑ Neutral ☐ Surprised

7. Dr. Wilson was there during his office hours.	☑ Neutral	☐ Surprised
8. You don't have my transcripts.	☐ Neutral	☑ Surprised
9. The T.A. was late again.	☑ Neutral	☐ Surprised
10. The class is cancelled.	☐ Neutral	☑ Surprised

EXAMPLE SCRIPT 14

CD 1, Track 15

Neutral: That's what he said.
Doubtful: That's what he said.

Neutral: I think so.
Doubtful: I think so.

Neutral: That's okay.
Doubtful: That's okay.

Neutral: The quiz is usually easy.
Doubtful: The quiz is usually easy.

Neutral: I could go to the health center.
Doubtful: I could go to the health center.

Neutral: You can turn your paper in late.
Doubtful: You can turn your paper in late.

Neutral: Joe would give me a ride.
Doubtful: Joe would give me a ride.

Neutral: I can do that.
Doubtful: I can do that.

Neutral: You could live in the dorm.
Doubtful: You could live in the dorm.

Neutral: I have one more excused absence.
Doubtful: I have one more excused absence.

PRACTICE 14

CD 1, Track 15 Resumed

1. Sometimes Professor Davis teaches the class online.	☑ Neutral	☐ Doubtful
2. You could use your credit card.	☐ Neutral	☑ Doubtful
3. The class might not be closed.	☑ Neutral	☐ Doubtful
4. My roommate said she'd pay half of the bill.	☑ Neutral	☐ Doubtful
5. You could ask for an excused absence.	☐ Neutral	☑ Doubtful
6. Your book could be at the lost and found.	☐ Neutral	☑ Doubtful
7. I can look on the Internet.	☑ Neutral	☐ Doubtful
8. We can take a bus.	☐ Neutral	☑ Doubtful
9. You could borrow Tom's notes.	☐ Neutral	☑ Doubtful
10. We could take a break.	☑ Neutral	☐ Doubtful

EXAMPLE SCRIPTS 15

CD 1, Track 16

Listen to part of a lecture in a linguistics class. The professor is talking about Gullah.

Example Script 1

So although many people, even those who live on the coastlines of Georgia, South Carolina, and Florida, believe that Gullah is a Native American language, it is really a creole that was common among the black populations in the seventeenth century. It's actually a mix of several African languages and the English that was spoken in the 1600s when British explorers and settlers were beginning to populate the southeast coast of what would become the United States. More than 300 years later, it is remarkable that approximately 7,000 people speak Gullah as their only language and another 250,000 use Gullah with friends and family at least part of the time! Most speakers live on the sea islands off shore, but some mainlanders also speak Gullah. The relative isolation of the islands, which were accessible only by boat until the twentieth century, has probably contributed to the perpetuation of the language.

Since American colonists had little experience with the cultivation of rice, they were willing to pay higher prices for slaves from the traditional rice region of Africa who were used to laboring in the rice fields. Consequently, a large number of the slaves were from Senegal, Sierra Leone, and Liberia. Recent research has revealed enough similarities between them for Gullah and Krio, which is a creole language spoken in Sierra Leone, to be mutually comprehensible. However, most of the vocabulary for Gullah is English, whereas the grammar and phonetic systems are more like those of West African languages. Most linguists agree that its origin is a practical response to the problem of communication among slaves from different tribes and countries with no common language. In fact—and this is interesting—it's speculated that the word *Gullah* may have come from the name of the Angolan tribe that was identified as "N'gola" or "Gullahs." Still in question is whether the language derived from West African Pidgin English, which the slaves may have been familiar with before they were brought to the Americas, or whether it sprang to life on the rice plantations of Georgia, South Carolina, and Florida. So that's a mystery for us to unravel! We really don't know . . .

Tone: Interested, enthusiastic

Example Script 2

So uh, although many people, even those who live on the coastlines of Georgia, South Carolina, and Florida, they uh, they believe that Gullah is a Native American language, it's probably a creole that was common among the black populations in the seventeenth century. It could be a . . . a mix of several African languages and the English that was spoken in the 1600s when British explorers and settlers were beginning to populate the southeast coast of what would become the United States. More than 300 years later, approximately 7,000 people speak Gullah as their only language and another 250,000 use Gullah with friends and family at least part of the time. Most speakers live on the sea islands off shore, but some mainlanders also, we think that they speak Gullah. The relative isolation of the islands, which were accessible only by boat until the twentieth century, has probably, probably . . . contributed to the perpetuation of the language.

Since American colonists had little experience with the cultivation of rice, they were willing to pay higher prices for slaves from the traditional rice region of Africa, who were used to laboring in the rice fields. Consequently, um . . . a large number of the slaves were from Senegal, Sierra Leone, and . . . and . . . Liberia. Recent research has revealed enough similarities between them for Gullah and Krio, which is a creole language spoken in Sierra Leone, for these languages to be mutually comprehensible. However, most of the vocabulary for Gullah is English, whereas the grammar and phonetic systems are more like those of West African languages. Well, most linguists agree that its origin is a practical response to the problem of communication among slaves from the different tribes and countries with no common language. In fact, it's speculated that the word *Gullah* may have come from the name of the Angolan tribe that was identified as "N'gola" or "Gullahs," but of course we can't be sure. Still in question is whether the language derived from West African Pidgin English, which the slaves may have been familiar with before they were brought to the Americas, or whether it sprang to life on the rice plantations of Georgia, South Carolina, and Florida. So, uh, we don't really know . . .

Tone: Reserved

Example Script 3

Professor: So although many people, even those who live on the coastlines of Georgia, South Carolina, and Florida, believe that Gullah is a Native American language, it is really a creole that was common among the black populations in the seventeenth century. It's actually a mix of several African languages and the English that was spoken in the 1600s when British explorers and settlers were beginning to populate the southeast coast of what would become the United States. More than 300 years later, it is remarkable that approximately 7,000 people speak Gullah as their only language and another 250,000 use Gullah with friends and family at least part of the time! So, what else have we learned about Gullah? Why is it still in use? Anyone?

Bill: Well, most speakers live on the sea islands off shore, but some mainlanders also speak Gullah, and the relative isolation of the islands, which were accessible only by boat until the twentieth century, well, that has probably contributed to the perpetuation of the language.

Professor: Good observation, Bill. Since American colonists had little experience with the cultivation of rice, they were willing to pay higher prices for slaves from the traditional rice region of Africa who were used to laboring in the rice fields. Consequently, a large number of the slaves were from Senegal, Sierra Leone, and Liberia. So . . . Yes, Jane?

Jane: I just wanted to mention that recent research has revealed enough similarities between them, for Gullah and Krio, which is a creole language spoken in Sierra Leone, for those languages to be mutually comprehensible. However, most of the vocabulary for Gullah is English, whereas the grammar and phonetic systems are more like those of West African languages.

Professor: Thanks for bringing that up. Most linguists agree that Gullah's origin is a practical response to the problem of communication among slaves from different tribes and countries with no common language. In fact, and this is interesting, it's speculated that the word *Gullah* may have come from the name of the Angolan tribe that was identified as "N'gola" or "Gullahs." Still in question is whether the language derived from West African Pidgin English, which the slaves may have been familiar with before they were brought to the Americas, or whether it sprang to life on the rice plantations of Georgia, South Carolina, and Florida. So that's a mystery for us to unravel! We really don't know . . .

Tone: Respectful, encouraging

PRACTICE 15

CD 1, Track 16 Resumed

Listen to part of a lecture in an astronomy class. The professor is talking about Saturn.

Although it was previously believed that Saturn had seven major rings and a few very faint minor rings, modern space telescopes are now able to detect tiny particles of dust and ice using infrared technology, and that has allowed us to view a new and larger ring—the eighth major ring actually—and it is huge! This gigantic ring circles Saturn from about 8 million miles away, so, uh, that means it is 50 times farther away than the other rings are, from Saturn, I mean. And, did I mention that it has a totally different angle from those of the other rings? It tilts 27 degrees from the planet's main ring plane. Another interesting aspect of the new ring is the size of the particles, which are so tiny, maybe 10 microns, if you can imagine that, and they're also so far apart that if, say, you were standing in the ring itself, well, you wouldn't even know it. In a cubic kilometer of space, for example, you probably wouldn't find more than twenty of these microscopic particles, maybe less. But, as I said, the ring is huge, so it's not only wide, it's also very thick, with a vertical height of, oh, about twenty times that of the planet itself! And at a temperature of about minus 316 degrees F, it's extremely cold. But the infrared telescopes are particularly well suited to identifying small, cold particles, and the eighth ring was found.

So where did this ring come from? Well, we think that it may have been part of one of Saturn's moons at one time, which may have broken off as a result of being barraged by comets and icy space rocks. That is, many small impacts could have swept dust into the atmosphere, which eventually migrated toward the planet and formed the ring.The moon, called Phoebe, orbits within the ring and in the same direction, which is important to note because the rest of the rings and most of the moons orbit in the opposite direction. And it is well documented that the cratered surface of the moon has been the target of many, many chunks of ice, space rocks, and other celestial bodies. Okay then . . .

Example Notes

Before—Saturn 7 major rings, few minor
Now—infrared technology—major ring 8
8 million mi/50 X farther than others
27° titl/ 10 micron particles/ 20 per cu k
20 X height planet
–316°F
Part moon Phoebe
Broken off/ impacts
Orbits same/ other rings, moons opposite

Tone: Interested

PRACTICE 16

CD 1, Track 17

1. Why don't you stop by my office at 2:00?
   Signal words: <u>Oh wait</u>.
   Reversal: Could you make that 2:30?

2. The last day to submit an application is Tuesday,
   Signal words: <u>I mean</u>,
   Reversal: Thursday.

3. Psychology 400 is closed.
   Signal word: <u>Oops</u>.
   Reversal: Not Psychology. Philosophy 400 is closed.

4. You can't check magazines and journals out of the library.
   Signal words: <u>Sorry</u>. <u>I meant</u>
   Reversal: new magazines and journals. Older periodicals can be checked out, though.

5. Sure, you can borrow my notes.
   Signal words: <u>Uh oh</u>.
   Reversal: I guess not. I don't have them with me.

PRACTICE 17

CD 1, Track 18

Listen to part of a lecture in a linguistics class. The professor is talking about the history of language.

Like families of people, languages are grouped into families, going backward in time on a genealogy chart. The Germanic family includes not only German but also English. And these two languages—English and German—they may be viewed as having a close family relationship like a brother and sister. Now French and Spanish are also related to English but not as closely as German. So French and Spanish are like brother and sister, but they have a relationship with English and German that would be more like that of cousins. As we go back in time, we see that French, Spanish, and of course several other of the languages in the Romance family claim Latin as a common ancestor. But let's go back even further—all the way back to tribes in Europe who lived there about six or seven thousand years ago. Linguists believe that these people spoke a language that was common to both the Germanic family and the Romance family—and that would be the Proto-Indo-European language, the oldest ancestor of both groups.

Now, as different as Germanic and Romance languages are, they do have some vocabulary in common from which hypothetical reconstructions of older languages have been made. Perhaps the most important breakthrough was made by Sir William Jones in the eighteenth century who found a connection between ancient Sanskrit, a classical Indian language, and the European languages. Okay then, with that piece of the puzzle, we have been able to put forward a fairly good linguistic analysis of the parent language, Indo-European. So the next question, of course, is this: Is it possible that all languages are related, beginning with a single ancestral language?

To investigate that fascinating possibility, linguists have studied and compared a large number of non-Indo-European languages in an effort to group them into families. For example, Sino-Tibetan, which includes Mandarin Chinese, also probably includes more than two hundred and fifty additional languages. The possibility of an ancestor that may be common for Sino-Tibetan and Indo-European, for example, is intriguing but difficult to prove. But some linguists believe that we can classify all of the languages into a small group of families that may have been related tens of thousands of years ago . . .

Topic:	The relationship between languages
References:	The references to brother and sister as well as to cousins
Reason for references:	Talking about family relationships may seem off topic, but the author is making a comparison between relationships among family members and relationships among languages.

PRACTICE 18

CD 1, Track 19

Conversation 1

Professor: I'm sorry that you are disappointed about the work-study position.
Student: Thanks.
Question: What does the man mean?
Function: Regret, but not apology

Conversation 2

Professor: All of Dr. Roger's sections are always filled the first day of registration.
Student: I know. I was lucky to get in.
Question: What does the man mean?
Function: Assumption. Dr. Roger must be a popular teacher.

Conversation 3

Student: Can you read my paper before I turn it in?
Professor: Sure.
Question: What does the woman mean?
Function: Request

Conversation 4

Student: Would you like to go out to eat?
Student: I have a presentation tomorrow.
Question: What does the woman mean?
Function: Refusal with an excuse

Conversation 5

Student: Jane is transferring at the end of the semester.
Student: Really? What makes you think that?
Question: What does the man mean?
Function: Disagreement or doubt

Conversation 6

Professor: You wanted to see me?
Student: Yes, Professor Henry. To tell the truth, we expected to get more points for our project. It took a lot of time to complete it.
Question: What does the man mean?
Function: Complaint

Conversation 7

Student: So you only got a C on the final.
Student: Yes, but I couldn't have studied any harder than I did.
Question: What does the woman mean?
Function: A strong defense

Conversation 8

Student: How about getting together to study for the test?
Student: Why not meet at the library in one of the study rooms?
Question: What does the woman mean?
Function: Suggestion

Conversation 9

Student: Guess what Professor Jones did in class today?
Student: What?
Question: What does the woman mean?
Function: Interest. To maintain attention

Conversation 10

Student: I could help you after class.
Student: It's really nice of you to offer, but I'd better do it myself.
Question: What does the woman mean?
Function: Polite refusal

PRACTICE 19

CD 1, Track 20

Listen to part of a lecture in an art history class. The professor is talking about Cubism.

Initially, *Cubism* was a rebellion by avant-garde artists in Europe against the previous artistic expressions, especially those by realists and impressionist painters who had appealed to the texture and color of the paint and the play of light in the canvas. Emerging in Paris about 1907, the year that Picasso and Braque were introduced, Cubism offered as an alternative to realism and classical forms of beauty a somewhat analytical and abstract system. Both Picasso and Braque had been influenced by Cezanne, a revolutionary artist who had gone well beyond the realism required of Impressionism by introducing a nontraditional role for perspective. His now famous declaration that everything in painting can be reduced to cubes, spheres, cylinders, and cones contributed fundamentally to the concept of Cubism that emerged as a joint effort between Picasso and Braque.

Cubists achieved fragmentation within a shallow plane or even several interlocking planes, intersecting in ways that removed the traditional depth and separation of background and subject. In effect, Cubists broke down their subjects into facets, reassembling them with sharp edges and angles, and showing several different views of one object at the same time. In this way, more than one perspective was represented in the same work and the subject was reconstructed from component parts. This style created a two-dimensional impression while at the same time revealing the front, back, and sides of the subject. Based on the premise that the eye is always moving, it was proposed as more real than realism because it is in fact the way that the world is perceived.

Textbooks often cite *Les Demoiselles d'Avignon* by Picasso in 1907 as the first Cubist painting, but an argument can also be made that Braque's series of landscapes, *L'Estaque,* the following year were more true to the Cubist tradition that would follow. The French art critic Louis Vauxcelles is thought to be the first to use the term *Cubism* after reviewing the Braque exhibits, referring to the geometric forms in the works as just a series of little *cubes*.

The movement prior to 1912 is generally considered Analytical Cubism because it was a very methodical process that focused on geometrical forms and muted colors. Some of the earlier paintings produced were almost monochromatic in brown, gray, or black. The palette was reduced in order to concentrate on the shapes and spatial construction. Among the most popular subjects were still lifes of the human face and form, musical instruments, and bottles or pitchers. Landscapes had almost disappeared as a theme for Cubists.

After 1912, Synthetic Cubism emerged, with more shapes, stencils, collages, and brighter colors. It was at this time that European artists were discovering the rich artistic traditions of Native Americans, Africans, and Pacific Islanders, and there is no doubt that they were influenced by the tribal designs and colors, which they appreciated, somewhat superficially for the stylistic rather than cultural or symbolic value. The Post Impressionist, Paul Gauguin, was already experimenting with vividly colored paintings and prints that he created while living in the Marquesas Islands.

So, although both Picasso and Braque remained at the center of the movement, other major proponents of Cubism included painters worldwide, including Juan Gris, Diego Rivera, and Max Weber. In addition, sculptors, notably Jacques Lipchitz, were producing work based on cubist forms. Moreover, the influence of Cubism can be found in some important literature of the era in which several viewpoints converge in a novel or a poem. For example, William Faulkner's novel *As I Lay Dying* captures the experiences of fifteen characters with diverse perspectives. Even some architects began to identify with Cubism, using multifaceted facades.

The Cubist school was largely disbanded after World War I, but the artistic departure from realism had a profound influence on the development of modern art. As photographic images that depicted reality with exacting clarity became more available in the twentieth century, artists attempted to represent reality in ways that could not be captured by the single perspective of the camera lens. Cubism provided an alternative view that was attractive as a point of departure for many modern forms of art, emerging as Vorticism in England, Constructivism in Russia, and Expressionism in Germany. Surrealism in Spain, and Pop Art in the United States also have their roots in the Cubist Movement . . .

Topic Sentence:
Initially, Cubism was a rebellion by avant-garde artists in Europe against the previous artistic expressions, especially those by realists and impressionist painters who had appealed to the texture and color of the paint and the play of light in the canvas.

Cues: Dates
1907
1912

Organization:
Chronology

PRACTICE 20

CD 1, Track 21

Listen to part of a lecture in an anthropology class. The professor is talking about evolution.

Paleontologists have identified two distinct human species during the last million years. *Homo erectus*, which is now extinct, and *Homo sapiens* from which all modern humans have evolved. But, if we evolved from *Homo sapiens*, then what happened to *Homo erectus*? Well, two theories have been proposed. The so-called *Multiregional* model, the *Multiregional* model, uh, which assumes that *Homo sapiens* evolved from *Homo erectus* and dispersed throughout the entire world, developing variations in major regions that formed the racial diversity in the world today. But I'm getting ahead of myself. About 1 million 700 thousand years ago, *Homo erectus* appeared in Africa and then spread throughout the world and evolved into regional variations, and then evolved into slightly varied local populations of *Homo sapiens*. For example, let's say, the Far Eastern region. Excavations have uncovered Chinese *Homo erectus* skulls with flattened faces and protruding cheekbones, similar characteristics found in modern Oriental populations. And it would have been from that regionally developed *Homo erectus* that modern *Homo sapiens* evolved in the Far Eastern region. This same history would have repeated itself throughout the various regions of the world with local variations that account for modern racial features. And clearly, this model supports the single, continuous evolution of the human species.

The other theory, referred to as the *Out of Africa* model argues that *Homo erectus* migrated out of Africa and spread throughout the regions of the world, and then, about 150,000 years ago, *Homo sapiens* developed in Africa and that *Homo sapiens* also moved out of Africa, also dispersing throughout the world. The *Out of Africa* model claims that non-African *Homo erectus* became extinct without evolving into *Homo sapiens* and, although interbreeding populations most certainly existed, *Homo erectus* was eventually displaced by *Homo sapiens*. A number of studies of human fossils from the past 50,000 years support this claim that many racial features have developed in the past 30,000 years, which calls into question the longer timeline proposed in the *Multiregional* model. And, oh, yes, it is worth mentioning that the *Out of Africa* model requires that we accept parallel development in the evolution of our species . . .

Example Notes

<u>Multiregional</u>
HS evolved from HE
HE dispersed world
w/ regional variations
1 m 700,000 yrs HE
HE—HS region

<u>Out of Africa</u>
HE Africa migrated
Spread world
150,000 years HS Africa—spread—displaced HE
Parallel development

Example Question

	Multiregional Model	Out of Africa Model
Continuous evolution of humans	✓	
Displacement of *Homo erectus* by *Homo sapiens*		✓
Separate migrations by *Homo erectus* and *Homo sapiens*		✓
Racial variation introduced by *Homo erectus*	✓	

CHAPTER 4—SPEAKING STRATEGIES

Example answers for the Speaking section provide only one possible response to a question or task. Use them to evaluate your answers. Remember that your answers may be different, but correct.

PRACTICE 1

CD 2, Track 1

1. Would it be okay if I speak English with some of the people here during the break?
2. I want to speak English during the break to warm up for the Speaking section.
3. Thank you. I appreciate it.
4. Thanks anyway. I'll just talk to myself.
5. Could you please tell me where the bathroom is?
   Could you please tell me where the water fountain is?

PRACTICE 2

CD 2, Track 2
Example Answer
San Diego is a city that I know well because I go there with my family every year. It's located in Southern California on the Pacific Ocean. The weather is warm all the time and the flowers are always in bloom along the highways. My family enjoys San Diego because there's so much to do there. We usually stay close to the beach so we can go swimming and surfing, but we also take advantage of the great shopping areas and restaurants downtown and across the bridge in Coronado. San Diego has a lot of parks, museums, and tourist attractions, for example, SeaWorld and the San Diego Zoo. Even though San Diego is one of the largest cities in the United States, it's divided into smaller districts like Old Town, Downtown, Balboa Park, and Mission Beach, and public transportation is fairly good so it's easy to get around the city.

Did you take deep breaths?
Did you smile before you started?
Did you speak up?

EXAMPLE SCRIPTS 3

CD 2, Track 3

Now listen to part of a lecture in a biology class. The professor is talking about hibernation.

To continue our discussion on hibernation, let's take a look at the hibernation routines of cold-blooded animals like frogs. Actually, cold-blooded animals have few internal resources to keep warmer or cooler than their surroundings because their body temperatures are controlled by the environment so few can survive long exposure to, let's say, more than a degree or two below freezing and at the other extreme about 100 degrees Fahrenheit. Okay then, in the summer when the temperatures soar above the 100-degree mark, what do cold-blooded animals do, like frogs for instance? Well, frogs often burrow themselves in the soil, especially in muddy stream beds. And in the winter, they hibernate in the earth under the frost line or under ice. In fact, it's hard to believe, but some frogs can survive during hibernation even when their bodies become frozen.

Now, you will recall that frogs have lungs, like we humans do. So how do they breathe under the earth? Well, in hibernation, their metabolism slows down until their hearts are beating at extremely long intervals and a very small amount of oxygen is required, so they can absorb the little bit of oxygen that they need from the water in moist soil or melting ice and they can do that through their skin. And that brings us to the problem of food, uh, nourishment. Frogs need only about 1 percent of the usual intake while they are hibernating. So if a frog hibernates for 100 days, let's say, and it doesn't eat at all, then it would really only be missing a day or two of the nourishment that it usually takes in during an active period . . .

Example Answer

Hibernation occurs in animals when the metabolism's reduced to a point of inactivity and body temperature and breathing are depressed. Hibernation can be total or intermittent with short periods of normal activity. Intermittent hibernation is also called "torpor." The lecture explains hibernation in frogs. They're cold-blooded animals so they hibernate during periods of extreme temperatures. In the summer, they bury themselves in the mud, and in the winter, they bury themselves under the ice. They can still breathe through their lungs because they don't need much oxygen during hibernation, and they can absorb it through their skin from the water in the mud or ice. And they eat very little—only about 1 percent of the nourishment that they usually require.

PRACTICE 3

CD 2, Track 3

Now listen to part of a lecture in a psychology class. The professor is talking about phobias.

There are over 700 different documented phobias, and about 10 percent of the adult population suffers from at least one. But today we're going to touch on two of the most common phobias. First on the list is arachnophobia, which is the fear of spiders. Now, most of us would prefer not to have a spider crawling on us, and there is a legitimate concern that a spider might bite, but I'm talking about a fear that is so intense that the people suffering from acute arachnophobia simply won't put themselves in a position in which they might *see* a spider. They won't go to a park, they won't eat a meal in the back yard, they won't take a walk in the neighborhood because a spider might be there.

Okay, what do you think might be second on the list? Acrophobia, which is the fear of heights. So people with this phobia may refuse to accept a job in places that might require them to see out a window on a higher floor of a building, and they won't go anywhere in a car if it requires driving across a high bridge, and in severe cases, they won't even climb a ladder. Some people are afraid they might fall, even if there is a safety barrier like a window or a rail. Other people say they are afraid that they might jump, but, of course, that is completely within their control.

So, I think you can see that phobias are grounded in some reasonable fear—the spider could bite, and the rail could give way—but the fear is exaggerated and it interferes with ordinary activities. That is what qualifies them as phobias . . .

Example Answer

A phobia's an unreasonable fear that interferes with a person's usual routine. There are more than 700 different phobias and they affect 10 percent of adults. The lecturer provides two common examples of phobias. The first is arachnophobia, which is the fear of spiders and the second is acrophobia or the fear of heights. It's important to remember that the fear's extreme and it causes people to avoid normal situations. People suffering from arachnophobia would refuse to go outside in the back yard or the neighborhood because they might see a spider, and people suffering from acrophobia would refuse opportunities that would necessitate their driving across a bridge or being near a window in an upper floor of a building.

PRACTICE 4

CD 2, Track 4

1. Volcanoes and geysers have some common elements.

   V + G common elements

2. Mercury is not a solid.

   M ≠ solid

3. Babies demonstrate understanding of mathematical concepts at about six months old.

   B understand M @ 6 mo

4. Few ancient cities are constructed without walls.

   few AC w/o walls

5. The paintings of flowers by O'Keeffe were larger than the works of her contemporaries.

   flowers O > contemporaries

6. The oxygen supply to the cornea of the eye comes directly from the air.

   oxygen C ← air

7. An example of a successful online business is eBay.

   X OB eBay

8. An ornithologist is a scientist who specializes in the study of birds.

   O = scientist birds

9. The architecture of many early public buildings includes a dome or a tower.

   A early PB dome o tower

10. Raindrops refract light like a prism, causing the colors in a rainbow.

    raindrops = prism → colors R

PRACTICE 5

CD 2, Track 5

1. Migration can cause variation in populations due to the introduction of new genes.

migration → variation ← new genes
pop

2. The right hemisphere of the brain controls the left side of the body and contributes to spatial, artistic, and musical tasks, whereas the left hemisphere controls the right side of the body and is essential to verbal and logical tasks.

brain

right hem. left hem.

left side body spatial artistic musical right side body verbal logical

3. Somatotyping is a system that describes personality in terms of an individual's physique.

somatotyping

↓

system personality
physique

4. Historians believe that the first coins may have been created about 700 B.C.E. in Anatolia; the Greeks and Persians were exchanging coins between 650 and 500 B.C.E.; by 300 B.C.E., the Romans had already begun to mint coins and distribute them at temples.

700 BCE Anatolia created

650-500 BCE Greeks, Persians traded

300 BCE Romans minted

5. The two major types of seed plants are gymnosperms and angiosperms.

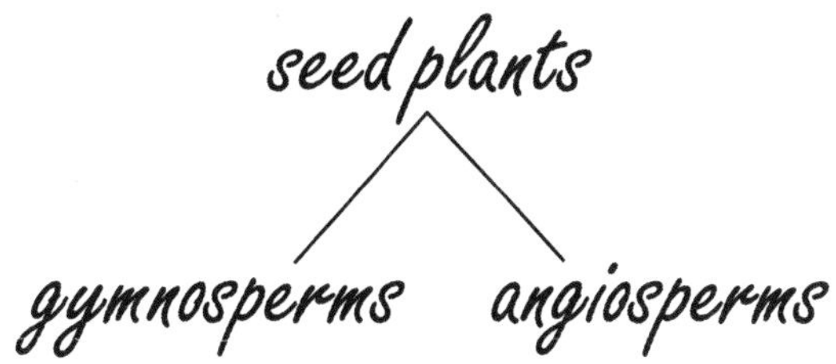

PRACTICE 6

CD 2, Track 6

Task 1
I would like to visit Los Angeles.

Task 2
I think that traditional schools offer many advantages.

Task 3
The student supports the new transfer program that the university has instituted.

Task 4
The lecturer's examples refute the research study on facial expressions.

Task 5
The woman's problem is that she has lost her student ID card.

Task 6
Three main types of volcanoes are identified in the lecture, including shield volcanoes, cinder cone volcanoes, and composite volcanoes.

PRACTICE 7

1. Some people think that employees should be paid based on achievements and merit. Other people think that seniority, that is, the length of time that they have been employed is more important. What do you think and why?

   Your opinion.

2. Explain the student's problem and the possible solutions that the professor recommends. What do you think that the student should do and why?

   First, the professor's suggestions. Then, your opinions.

3. Using the ideas and examples from the lecture, explain the three types of nonverbal behavior that the professor mentions in the lecture.

   Facts without an opinion.

4. The man expresses his opinion about the professor's policy for class participation. State his opinion and the reasons he has for having that opinion.

   The man's opinion.

5. Describe an ideal job. Use specific reasons and details to support your response.

   Your opinion.

6. Explain how the research study that the professor cites in his lecture supports the use of clean coal as an alternative energy source.

   Facts without an opinion.

7. Explain the woman's problem and the two possible solutions that her friend suggests. Then choose the better option and justify your choice.

   First, her friend's suggestions. Then your opinion.

8. Some people believe that it is better to marry young. Other people advise young people to wait until later in life to marry. Which idea do you think is better?

   Your opinion.

9. Using the main ideas and supporting examples from the lecture, explain how laser beams work and how they support modern products and technologies.

   Facts without an opinion.

10. The man expresses his opinion about the requirement that his advisor sign all of his course requests before registration. State his opinion and the reasons that he gives for holding that view.

    The man's opinion.

PRACTICE 8

CD 2, Track 7

1. The <u>lecturer</u> describes two types of competition.
2. I <u>believe</u> that the man should join a study group.
3. A number of <u>variables</u> should be taken into consideration.
4. <u>For one thing</u>, the parking lots are always full.
5. In my <u>view</u>, students should live on campus.

6. I'd rather go to school at night than during the day.
7. Virtually every student signed the petition.
8. Personally, I think that the steam engine was the most important invention.
9. The professor makes a very good case for changing the schedule.
10. I think that grades should be based on tests, not on class discussions.
11. The reading passage summarizes the theory of plate tectonics.
12. It's a valid argument.
13. Third, exercise is essential.
14. The example in the talk clearly explains the concept.
15. The man's friend recommends that he take a break.
16. The research was verified by a second study.
17. The lecturer disagreed with the method.
18. Generally speaking, mathematics is the study of time, quantity, and distance.
19. In conclusion, insects adapt to survive.
20. Several thoughts were presented.

PRACTICE 9

CD 2, Track 8

1. Although________________, I prefer________________________________.
   Although many students enjoy living in a dormitory, I prefer to live in an apartment.

2. Even though_____________, I prefer________________________________.
   Even though tours are more convenient in many ways, I prefer to travel with a friend.

3. Despite ________________, I prefer________________________________.
   Despite the benefits of working for a large company, I prefer to start my own business.

4. In spite of _______________, I prefer_______________________________.
   In spite of the advantages of e-mail and texting, I prefer to conduct important communications by telephone or in person.

5. ________________________, but I prefer____________________________.
   Living alone is probably easier in many ways, but I prefer to live with a roommate.

6. ________________________; however, I prefer________________________.
   It can be argued that you should never tell a lie; however, I prefer not to hurt a friend's feelings when the truth is harmful.

7. ________________________; nevertheless, I prefer____________________.
   I enjoy cultural activities; nevertheless, I prefer to go to sports events.

8. Although________________, I would rather___________________________.
   Although some people like to see serious dramas, I would rather see comedies when I choose a movie.

9. In spite of_______________, I would rather___________________________.
   In spite of the time that is required to participate in campus activities, I would rather join a student club to make friends.

10. ________________________, but I would rather_______________________.
    Some people like to have a lot of friends, but I would rather have a few good friends.

PRACTICE 10

1. What is your favorite place to go when you have free time?
   When I have free time, I like to go to take a walk because it is good exercise and it gives me a chance to explore the area.

2. Do you think that it is better for a family to live in one place or move to different places?
   I think it is better for a family to live in one place because they can maintain close relationships with family, friends, and neighbors.

3. In your opinion, what is the most important skill or lesson you learn in college?
   I think that learning to live independently is the most important lesson in college because it involves both professional and personal skills.

4. Some people believe that they can succeed by following a plan while other people believe that they will succeed by taking advantage of chance opportunities. What is your opinion?
   I believe that success requires a plan because it assures that you will make progress toward a goal, whereas chance opportunities are random and uncertain.

5. Who is your most important advisor?
   My most important advisor is my mother because she is very wise and she has my best interest in mind when she gives her opinion.

6. The university can use a grant to build a sports center or a branch library. Which project do you favor?
   I would prefer that the university use their grant to build a branch library because the main mission of the school is to provide educational programs and support.

7. Would you rather own a business or work for a company?
   Eventually I would like to own my own business because small business owners can control many aspects of their lives, including where they work, the number of hours that they are on the job, and to a certain extent, the amount of money that they earn.

8. Should men and women attend the same school or separate schools?
   I think that men and women should attend the same schools because they need to learn how to socialize and how to compete.

9. If you received a gift of one thousand dollars, what would you do?
   If I received a gift of one thousand dollars, I would use it to pay the balance on my credit card because interest on an unpaid balance is very high.

10. Some people like to make new friends, and other people prefer to spend time with friends that they have known for a long time. What do you prefer to do?
    I prefer to spend time with friends that I have known for a long time because I feel comfortable with them and we will probably continue our relationship over our lifetimes.

PRACTICE 11

1. According to the reading passage, the jaw of the great white shark is lined with rows of replacement teeth.

2. According to the lecturer, only 2 percent of Antarctica is free of ice.

3. According to the professor, secondary colors are made by combining two primary colors.

4. According to the reading, estuaries are habitats for diverse wildlife.

5. According to the study, the blue whale can dive approximately 100 feet, which is about 250 meters.

6. Anthropologist George Wharton James concludes that the Hopi people have traditionally made pottery and woven baskets. James points out that bright primary colors such as red, green, and yellow are typical of the designs. He mentions that either geometric forms or ceremonial Kachina figures dominate the patterns.

7. Professor Abdul Latiff Mohammad states that the *Rafflesia* is a parasitic flowering plant found in the rainforests of Indonesia, Malaysia, and the Philippines. Mohammad maintains that it takes some species as long as six to nine months to complete the budding period. He concludes that a single bloom of the *Rafflesia* plant can grow to more than 3 feet long and weigh over 15 pounds.

8. Dr. William Fry proposes that laughing seems to provide the same benefits as exercise. Fry argues that laughter increases the release of endorphins in the body. He points out that when patients laugh, they require less pain medication.

9. Dr. Carl Sagan maintains that life on Earth is not exceptional. Sagan suggests that life probably exists elsewhere in such a vast universe. He proposes that consistent physical laws increase the probability of finding extraterrestrial life.

10. Dr. Edward Hall proposes that 65 percent of the meaning in a normal conversation is transmitted through nonverbal cues. Hall maintains that people from different cultures use gestures in different ways. He points out that sometimes a harmless gesture in one culture is interpreted as offensive in another culture.

PRACTICE 12

CD 2, Track 9

Task 1

Talk about your favorite subject in school. Include specific reasons and details to explain your choice.

Preparation Notes

Favorite subject

Father advised English—favorite

Reasons

- Good languages—learned quickly/best student
- Good teacher—music, culture, games/fun
- Useful—many people, Internet

Example Answer

My father advised me to take English in school. Even though I didn't want to at the time, I'm glad I did. It soon became my favorite subject for a couple of reasons. First, because I was good at languages, and I learned English quickly. So I was the best student in class, and that encouraged me, and made me like the subject even more. Second, I had a good teacher. She brought music and culture and, uh, games into the classroom so it made it fun. And last, it became my favorite class because I soon realized how useful it was. Many people speak English throughout the world, and now English is used on the Internet. I'm glad I took my Dad's advice and took the class.

Task 2

Some people like to be engaged in several tasks at the same time. Other people prefer to focus on one task at a time. Which do you prefer, and why? Use specific reasons and examples to support your opinion.

Preparation Notes

Concession

Multi task, many enjoy—I get confused

Choice

Focus one task

Reasons

- More efficient—faster
- Motivating—sense accomplishment

Conclusion

Efficiency, performance, motivation, energy

Example Answer

Although multitasking is considered an asset and many people seem to enjoy it, I prefer to focus on one task at a time. When I'm going back and forth between projects, sometimes I get confused and it takes time to sort it out. But by working on one task, I'm more efficient and I can complete it faster than I would by dividing my time. In my experience, I also do a better job. And I find that completing a task is very motivating. I feel a sense of accomplishment about ending a project, and I'm more excited about starting something new. So for efficiency, performance, motivation, and, uh, energy, I need to work on one task at a time.

Task 3

Read an announcement from the campus newspaper about a new program at the university.

Reading Time: 45 seconds
Refer to page 145.

Now listen to two students who are talking about the new Child Development Center at the university. The woman expresses her opinion of the program. Report her opinion and explain the reasons that she has for having that opinion.

Student 1:	I suppose you're happy about the new Child Development Center.
Student 2:	I sure am. I've already been over to see it, and I enrolled my daughter in the fall program.
Student 1:	How old is she now?
Student 2:	Two. She's just the right age to benefit from the preschool curriculum.
Student 1:	So how will that work for you?
Student 2:	I'll just drop her off in the morning when I come to campus, and then while I'm in class, she'll be at the center. If my classes are over before noon then I'll spend some time studying in the library.
Student 1:	That sounds good.
Student 2:	We've needed something like this for a long time. But really, I can't even begin to tell you what a good job they have done. The environment is child friendly and inviting, and everywhere you look there is another opportunity to learn and explore. I was so impressed. And the price is right.
Student 1:	Really? I would have thought that it would be expensive. I think the facility cost 12 million dollars to complete.
Student 2:	True. But it serves college students in the Education program as well. So their tuition also supports the center.
Student 1:	Oh.
Student 2:	It's a great opportunity for the students in the Education program to gain experience working with children. So, everybody wins.

<u>**Preparation Notes**</u>

Announcement

Child Development Center
12 m—lab Early Child/150 preschool

Opinion
Happy—enrolled daughter/2

Reasons
- Convenient—morning—class/library
- Ex. Environment—explore
- Reasonable price—2 functions

Example Answer
The university's opened a new child development center which will serve both as a lab for students in the College of Education who are pursuing a degree in early childhood education and also as a school for 150 preschoolers. The woman's enrolled her two-year-old daughter in the morning program. She's happy because the location's convenient, the center offers a good environment and curriculum, and it's not expensive. It's convenient for them because the woman and her daughter can ride to school together. Then she can attend her classes at the university and study in the library while her child's in the program. She says that it's an excellent environment for learning and exploration at a reasonable price, right on campus. Because the center serves both university students and children, there are two sources of tuition and the price is lower than would be expected at a 12-million-dollar facility.

Task 4
Read the passage about nonverbal communication.

Reading Time: 45 seconds
Refer to page 146.

Now listen to part of a lecture in a linguistics class. The professor is talking about nonverbal communication. Using information from both the reading passage and the lecture, discuss nonverbal communication and provide examples from the research study.

Drawing on the work of Dr. Richard Frankel, a research study was conducted by the College of Medicine in cooperation with the Linguistics Department. Patient-doctor communication was studied with a view to identifying mismatches between the verbal histories that patients provided and the nonverbal communication that accompanied them. Using videotaped sessions, researchers concluded that the verbal report patients gave their doctors often conveyed different information from their nonverbal signals.

Interestingly, preconceptions about detecting a mismatch didn't seem to apply. For example, patients who avoided eye contact and shifted in their seats were not necessarily those who gave inaccurate verbal reports. Typically, patients whose verbal and nonverbal cues were judged to mismatch blinked their eyes and nervously touched themselves more than normal. It was also found that facial expressions that were inconsistent with movement in the lower body suggested a mismatch in messages. For example, a patient who was smiling while nervously swinging his feet often did not give the doctor an accurate verbal report.

As a result of these studies, training at the College of Medicine now routinely includes nonverbal communication. Medical personnel are alerted to the problem of mismatched speech and nonverbal behavior. They are also reminded that most of a communication is transmitted by nonverbal cues. So watching what a patient does is often more important than listening to the history.

Preparation Notes

Reading Concept
Nonverbal—70–90%/facial, eye, gestures, distance, tone

Transition

Lecture Examples
Frankel—patient-doctor mismatches/verbal history, nonverbal behavior
- Blinking
- Touching themselves
- Conflicting facial—nervous movement legs

Conclusion
Training nonverbal—watching more than listening

Example Answer
Studies indicate that between 70 and 93 percent of any communication is conveyed by nonverbal behavior such as facial expressions, eye contact, gestures, distance, and tone of voice. Using Frankel's work as a basis for the study, researchers observed patient-doctor communication in an effort to identify mismatches between verbal histories and nonverbal behavior. Blinking and touching themselves were identified as signals that a mismatch was occurring. Conflicting facial expressions and nervous movement of the legs also indicated a mismatch. Training of medical personnel now includes nonverbal cues. Watching patients can be more significant to a diagnosis than listening to the history.

Task 5
Now listen to a short conversation between a student and his friend. Describe the man's problem and the solutions that his friend suggests. What do you think the man should do, and why?

Friend: Joe, are you okay?

Joe: Just tired. I can't sleep. The walls in my apartment building are like paper, and my neighbors are up until three in the morning—every morning.

Friend: Wow. And you have early classes, too.

Joe: Tell me about it.

Friend: I suppose you've already talked with the apartment manager.

Joe: Several times. And she's always very nice, and she says that she brings them into the office and talks with them about the noise.

Friend: But nothing happens?

Joe: Nope.

Friend: Well then, have you talked with some of your other neighbors? I mean, you can't be the only one who has a problem with these people.

Joe: I thought about that, but my neighbors are all students like me, and our schedules are really busy, with classes, part-time jobs and I hardly ever see them around.

Friend: You could leave them notes though. And then you could all complain to the manager. That should get her attention.

Joe:	It's a good idea, but I still think it would be hard to coordinate it.
Friend:	Okay. But you have to do something. Maybe you should just move.
Joe:	I have a lease that isn't up until the end of June.
Friend:	No. I don't mean you should move out of the building. I was thinking of moving to another apartment in the same building. That way you wouldn't be breaking your lease, but you could move away from the noise.
Joe:	Oh, I see what you're saying. That could work. I'm so swamped with school, though. It would probably take a whole weekend to get everything packed and moved and unpacked.
Friend:	You don't have that much stuff. I'd help you, and so would some of your other friends.

Preparation Notes

Problem
Noisy neighbors—3 am
Manager—not resolved

Suggestions
- Other neighbors—in person/notes
- Move—same complex

Opinion
Move—check neighbors first

Reasons
- Make sure quiet
- Solve in one weekend

Example Answer
Joe's problem is that his neighbors keep him up until three in the morning and then he has to get up early for classes. He's already spoken with the manager of the apartment building. But, uh, when she talks with the neighbors, the problem's not resolved. Joe's friend thinks he should get in touch with his other neighbors, either in person or by leaving them notes. They could complain as a group to the manager. Um . . . the other option she mentions is that he could move to another apartment within the same apartment building. She offers to help him move. So . . . in my opinion, the man should just move to another apartment in the complex, but first he should talk with the neighbors near the new apartment to make sure that it's a quiet place. That way, he won't trade one noisy neighbor for another. If it only takes him a weekend to solve his problem, it's worth it.

Task 6

Now listen to part of a lecture in an environmental science class. Using the main points and examples from the lecture, discuss research in the canopy of the rainforest.

Traditional ecological studies of the rainforest had been limited to species that occupied the forest floor. Many species of trees had been identified as well as plants and animals that spent most of their lives at ground level. But recently, some ecologists have begun to explore the canopy, that is, the tops of trees, and the ecosystems that contain species seldom seen on the forest floor.

Of course, the idea that species might live out their existence on a level above that which was the focus of scientific study—that was easy to accept. But how to approach the study of these species was a little harder to figure out. At first, we tried to use techniques and equipment adapted from mountain climbing, but that was often very dangerous, and for some researchers, impossible to accomplish. Besides that, the area of view was relatively restricted and did not allow us to follow animals very well as they traveled though the canopy. Then we started bringing in cranes and big construction equipment, which allowed researchers to be transported high into the treetops. And canopy observation platforms with walkways between them made the investigations safer and opened up the field to researchers who weren't able to use the mountain climbing gear. The platforms now swing above the canopy without disturbing it, allowing us to observe the ecosystem from above.

And we are finding that the rainforest canopy is one of the most diverse biosystems in the world, including between 50 and 80 percent of terrestrial species, depending on estimates. Besides the support trees and vines, many plants and arboreal mammals and reptiles, birds, and bats are found along with uncounted species of insects, spiders, and other arthropods. Now it seems that new species are being discovered daily, and not one at a time, but by the hundreds.

<u>**Preparation Notes**</u>

Main Idea 1

Traditionally—ground level

Main Idea 2

Lately—canopy

- Mountain climbing gear—dangerous, challenging
- Construction equipment—platforms, walkways

Main Idea 3

Broader area

- 100s species
- 50–80% canopy—diverse biosystem

Example Answer

Traditionally, researchers studied the plants and animals at ground level in the rainforest, but lately, ecologists have started to explore ecosystems in the canopy of trees. To do that, at first they used mountain climbing gear to get to the top, but it was dangerous and physically challenging for some of the researchers. Later, construction equipment was used to lift researchers into the canopy. By building platforms and walkways, it's now possible to observe a broader area so researchers can watch animals move around in their habitat. Now that researchers are investigating the rainforest from bottom to top, they're discovering hundreds of new species of plants, mammals, reptiles, birds, and insects. It's estimated that 50 to 80 percent of the species can be found in the rainforest canopy. The rainforest canopy is one of the most diverse of all the biosystems on Earth.

PRACTICE 13

CD 2, Track 10

Task 1

My husband is my favorite person.

My hometown is the place that I like to visit most.

person
object
event
activity

In my experience, watching TV is a good way to learn English.

In my view, the National Geographic channel is one of the best choices.

If I could change one thing about the world, it would be water pollution.

Task 2

Although eating at a restaurant is fun, I prefer to prepare most of my meals at home.

Although a case could be made for discussion classes, I like lecture classes.

In my experience, playing sports is important because it teaches children many lessons.

I think that working inside is better than working outside because it is more comfortable.

I agree that education should be free because it gives everyone the chance to find a good job.

I prefer the Internet to the library because I have access to more research.

Task 3

According to the announcement, the university is going to build married student housing.

The man objects to the policy because it is expensive.

He favors the current plan.

He presents three arguments against the proposal.

The woman agrees that parking is a problem.

The woman agrees with the policy. She explains that assigning spaces for visitors would help.

The woman thinks that charging a parking fee is a good idea. She points out that visitors should not use student spaces.

Task 4

An herbivore is an animal that eats plants.

According to the reading passage, art should not be censored.

The lecturer refutes the information in the reading.

Task 5

The problem is that the man's advisor is taking a job at another university.

According to the woman, one solution is to ask the advisor to recommend a replacement.

Another possibility is to make an appointment with the department chair.

I think that the best solution is to talk with his advisor because the replacement will probably agree with the advisor's previous plan for the student.

In my opinion, the man should ask his advisor for a recommendation.

Task 6

According to the lecturer, workers in the U.S. have about 15 paid vacation days per year.

The lecturer states that the number of vacation days is lower than that of most countries.

The two/major problems are health and family relationships.

The lecturer concludes that the quality of life suffers when workers do not take vacations.

In other words, time away from work is necessary for individuals and societies.

PRACTICE 14

CD 2, Track 11

Examples

1. Natural, spoken English: If I could meet anyone, I'd like to [liketa] meet Bill Gates.
2. Natural, spoken English: The University is going to [gonna] change its policy.
3. Natural, spoken English: In the first place, it's important for teachers to set a good example.
4. Natural, spoken English: The lecturer explained why the experiment didn't work.
5. Natural, spoken English: I think that the man ought to [oughta] talk to his professor.
6. Natural, spoken English: The lecturer claims that his research doesn't support the theory.
7. Natural, spoken English: If visitors came to my hometown, I'd take them to the beach.
8. Natural, spoken English: According to the reading passage, glaciers could have [coulda] caused the unusual topographical features.
9. Natural, spoken English: I admire him because he's a very honest person.
10. Natural, spoken English: According to the professor, they've formed over millions of years.

Practice

1. Natural, spoken English: It's clear that suburbs have been an important part of urban development since the 1920s.
2. Natural, spoken English: If the university increases the fees, students will have to [haveta] pay retroactively.
3. Natural, spoken English: The woman's problem is that she's not eligible for a scholarship because she's an out-of-state student.
4. Natural, spoken English: The woman suggests that the man register for a class that won't conflict with his lab.
5. Natural, spoken English: The man's friend thinks that he should have [shoulda] taken the work-study position.
6. Natural, spoken English: I don't agree with the idea for three reasons.
7. Natural, spoken English: I think that the woman should apply for an internship because she'll gain valuable experience.
8. Natural, spoken English: According to the lecturer, there might have [mighta] been water on other planets.
9. Natural, spoken English: In the lecture on satire, the professor states that it's been used in both visual and written works.
10. Natural, spoken English: A lot of [alotta] people are participating in the research studies.

PRACTICE 15

1. According to the professor, the study was real successful.

   According to the professor, the study was very successful.

2. The speaker claimed that the adjustment in fees was going to freak out a bunch of students who had not planned for an increase.

   The speaker claimed that the adjustment in fees was going to upset a number of students who had not planned for an increase.

3. After class, I usually hang out with some guys from my dorm.

   After class, I usually spend time with some friends from my dorm.

4. The lecturer said that kids who have asthma may experience a spontaneous cure when they grow up because their immune system improves in adulthood.

   The lecturer said that children who have asthma may experience a spontaneous cure when they grow up because their immune system improves in adulthood.

5. When I go back to my country, I will have to take a lot of stuff for my family and friends.

   When I go back to my country, I will have to take a lot of things for my family and friends.

6. In my opinion, like the best way to prepare for the TOEFL is to take practice tests.

   In my opinion, the best way to prepare for the TOEFL is to take practice tests.

7. I have had several cool teachers, but the best was Mr. Young because he was very knowledgeable and he made learning fun.

   I have had several good teachers, but the best was Mr. Young because he was very knowledgeable and he made learning fun.

8. The man's problem was that he messed up about the date of the field trip and he scheduled two important meetings with his study group in another class.

   The man's problem was that he made a mistake about the date of the field trip and he scheduled two important meetings with his study group in another class.

9. I agree that attending an Ivy League school is totally awesome.

   I agree that attending an Ivy League school is totally impressive.

10. A bunch of students signed a petition to change the payment policy for dormitory residents.

    A group of students signed a petition to change the payment policy for dormitory residents.

PRACTICE 16

CD 2, Track 12

Script

In the professor's opinion, the term *Dark Ages* is inappropriate and unfair. Modern scholars prefer the term *Early Middle Ages* to refer to the period between 400 and 1,000 C.E. He claims that merchants and crusaders brought culture and knowledge from the East to Europe. For example, Islamic education was influential when the first medieval universities were established in Europe. He also mentions that books were available and literacy rates had gone up from previous times. Art, architecture, and music began to develop along classical lines. Although he admits that the decline of the Roman Empire resulted in wars and political disorganization, he says that the political and economic systems were already reorganizing around a feudal system. He thinks it's unfair to compare the Early Middle Ages to later periods in history like the Enlightenment. Within the historical context, the Early Middle Ages includes many achievements. It's not the *period* that is dark. It is our *lack of information* about it . . .

Words or Phrases

term DA inappropriate
Modern Early Middle Ages 400–1,000 C.E.
merchants, crusaders East ➔ Europe
Islamic ed ➔ universities
+ books, + literacy
classical art, arch, music
decline R Empire ➔ wars, political disorganization
feudal
unfair comparison EMA—Enlightenment
EMA achievements
lack information, not period dark

PRACTICE 17

CD 2, Track 13

Notes—Contradiction

Summary of announcement
research study psych dept @ relationships undergrads 18–21 full time $20 1 ½ hr organization/ 100 ? survey

↓

Transition
woman yes/man no

↓

Man's opinion
use money but no

↓

Reason
time unrealistic disorganized/all afternoon—friends

↓

Reason
W says anonymous M doesn't want ?? personal life

Example Answer

A research study at the Psychology Department will investigate relationships among undergraduate students in college. To qualify, participants must be 18 to 21 years old and they must attend the university full time. The compensation is $20 for spending about an hour and a half with the researchers, including time for organization and responding to a 100-question survey. The woman's going to do it, but the man isn't. The man says he could use the money, but he doesn't plan to participate because he thinks that the time estimated is unrealistic. He is concerned that it'll be very disorganized and the participants will actually spend all afternoon there like some of his friends did for another experiment. He doesn't have time for it and, uh, the woman says the survey will be anonymous, but he doesn't want to share information about his personal life.

Audio CD 2, Track 13

Notes—Extension

Summary of concept

E interdisciplinary
equipment + human = safety/productivity

↓

Transition

case Billing Plus demonstrated importance

↓

Example

computer injuries
E training
equipment/exercise/breaks
keyboarding 5 hrs/day

3 yr/claims—50%
—absences
—budget for temps

↓

Relationship

E improved safety/productivity

Example Answer

Ergonomics is an interdisciplinary study of the way that equipment and human beings interact in the workplace with a view to improving safety and productivity. The case study of "Billing Plus" in the lecture demonstrated the importance of ergonomics to employees. When an increase in office injuries was attributed to computer use, the company established an ergonomics training program, purchased and adjusted equipment, recommended exercises and breaks, and limited keyboarding to five hours per day. Over a three-year period, claims decreased by 50 percent, and the company saved two dollars for every dollar spent on the program. A lot of the savings was the result of fewer absences and a smaller budget for temporary employees. So, uh, we see how ergonomics improved safety and productivity in this case.

PRACTICE 18

✔ 1. Finally, let me reiterate the qualities that I require in a friend—loyalty, dependability, and compatibility. And that's why I like Sarah so much.

2. In closing, let me say thank you for giving me your time today. I know that you are very busy.

   Although this is a good way to end a speech, it is not appropriate for the TOEFL.

✔ 3. In conclusion, all three types of dinosaurs were common, but they probably did not live during the same time period.

4. So that is my opinion, and I hope that you understand my English.

   You should not make an excuse for your English at the end of a speaking task.

✔ 5. And that's why I prefer public transportation—it's cheaper and it decreases the amount of pollution in the environment.

6. That's all. I'm sorry that I can't think of anything else.

   You should not make an apology at the end of a speaking task.

✔ 7. So the student disagrees with the university's new policy because it will be inconvenient for students and expensive in the long run.

✔ 8. Okay then. We see from the lecture that quilts were a good example of American folk art.

✔ 9. So I think that the man should talk with his professor before he makes a decision, because, if the professor says "no," then he can always try the other option.

10. I don't have time to say more. Thank you.

   Do not mention that you have run out of time at the end of a speaking task.

PRACTICE 19

CD 2, Track 14

Listen to part of a lecture in a psychology class. The professor is talking about learning.

Students show preferences for the way that they process information, either by seeing or hearing, logic or intuition, memorizing, visualizing, or exploring. Although there are a number of models to explain and exploit these preferences, one of the most common is that of Fleming. Fleming actually divides learning styles into four categories, but today we'll discuss the two most common ones for college students—a visual learning style and a verbal learning style.

If students prefer a *visual* learning style, they would rather see images, pictures, diagrams, charts, and maps in order to organize their information. People who depend on this style tend to have good spatial skills, and that usually translates into a good sense of direction. So, if this is a student's learning style, that student will be depending on the teacher to use the whiteboard or overhead slides to refer to visuals in order to explain the lesson. Handouts with images support students with a visual learning style. If those images aren't a part of the curriculum, visual learners can get distracted or confused by all the verbal input. They may even start doodling on their notes!

On the other hand, if students prefer a *verbal* style, they are perfectly happy with the written and spoken word. Reading the textbook and listening to the lecture are fine for them. If this is a student's learning style, that student probably has a good vocabulary and good note taking skills. They will be happier in a class where there is an opportunity to listen and discuss the information. Small groups are attractive to verbal learners because they can use their verbal skills to communicate with others.

But here's the problem. Often a mismatch can occur between the learning style and the teaching method. A good student in one class can be bored and inattentive in another . . .

Question:
Using the main points and examples from the lecture, discuss the two different learning styles that the professor explains.

Audio CD 2, Track 14
Response 1
Students like learning some ways more better than other ways. Visual and verbal learning are two, and they are learning styles that lecturer discussed. Um, and so students who they are visual learners, uh, visual learners, uh, they like pictures and maps and charts, and they pay attention when teacher uses blackboard or slides. Verbal learners like lectures and textbooks, and they talk a lot in class and participation more. Problem is that teaching method and learning styles may not be the same like they are, and then students get bored of it when method in class not matching their learning styles.

Level: 2.5
Rater's Comments:
The speaker responds to the question with accurate content. He also demonstrates the ability to maintain a continuous pace with some hesitations. Some limitations in grammar, vocabulary, and pronunciation interfere slightly with comprehension, but, in general, the talk is logically organized and the speaker is able to express his ideas.

Response 2
Students tend to prefer certain learning styles. In the lecture, two styles are discussed—the visual and the verbal styles. Students who like to learn visual benefit to pictures, maps, charts, and other images. If the teacher uses the board or slides to present material and gives handouts with visuals on them, then visual learners will be successful. But, students who like to learn verbal will pay attention to a lecture and depend on the textbook. They'll also talk in class and participate in small group activities. So, if the teacher relies to lecturing and also provides communication opportunities, then verbal learners will be more successful. The problem that learners have is when the teaching method and their learning styles different. Then, students who are high achievers in a class that supports their learning style can get bored when a teacher uses a method that it doesn't support their learning style.

Level: 3.5
Rater's Comments:
The speaker responds effectively to the topic, the content is accurate, the point of view is clear and well developed with examples and details. For the most part, the pace is continuous with very few hesitations. Occasional and minor errors in pronunciation and grammar are noted, but they do not interfere with the listener's ability to understand the communication. Overall, this is an excellent response.

PRACTICE 20

CD 2, Track 15

Task 1
I think the greatest invention in the world is the wheel because it's an ancient invention that continues to be used into modern times, and that in itself is a tribute to its utility. Of course, the wheel revolutionized the transportation system—from one stone wheel to four wheels on wooden carts to the wheels on jets and space vehicles. We're still using wheels to move heavy loads and people efficiently. But wheels are also an important part of many, uh, machines in the form of gears, rotors, and other necessary parts. Even the computer disc is a wheel of sorts. So the wheel's a great invention by itself or as an invisible but important part of a lot of other inventions.

Strategies:
2 Stay positive and confident—strong voice
6 Use a direct approach—no lengthy introduction here
7 Give your opinion if you are asked for it---opinion required here
8 Be sure of your choices—vocabulary and grammar are familiar to the speaker
10 Support your answers with reasons
12 Use the directions as an outline—45 seconds for opinion and two reasons
14 Use written and spoken contractions—written contractions throughout
15 Use colloquial vocabulary—no slang here
18 Stop when you have finished answering the question—good summary conclusion

Task 2
It's fun going to sports events, but I prefer attending cultural activities. First of all, I'd much rather play sports than watch. When I'm not in the game, I find my mind wandering and I'm looking for something to do. Very often, I drink too much or eat too much at sports events. On the other hand, when I attend a cultural activity like a play or a concert, I'm caught up in the music or the plot. I feel mentally engaged without being a participant. I'm happy to be entertained, and I don't need beverages or snacks to have a good time. Since I'm naturally more interested in the fine arts, I probably understand cultural activities better than sports, and that adds to the enjoyment.

Strategies

2 Stay positive and confident—strong voice
7 Give your opinion if you are asked for it—opinion required here
8 Be sure of your choices—vocabulary and grammar are familiar to the speaker
9 Acknowledge the other opinion—"It's fun going to sports events, but . . . "
10 Support your answers with reasons
12 Use the directions as an outline—45 seconds for opinion and two reasons
14 Use written and spoken contractions—written contractions throughout
15 Use colloquial vocabulary—no slang here
18 Stop when you have finished answering the question

Task 3

The Tutoring Center's advertising for tutors in freshman math courses. The tutors will help groups of students in Math 100. To qualify, tutors must have received a B or better in Math 100, and they need to have good communication skills. The man qualifies for the position and has experience as a tutor, but he's not going to apply because he's had a bad experience working at the Tutoring Center. Um . . . he says his student was always late, and he wasn't paid extra for the time he spent waiting or for the additional work that was required to go over the student's homework. When he calculated his compensation, he made less than five dollars an hour, and he doesn't want to repeat that experience. The woman tries to convince him that it would be better tutoring a group, but he says he's not excited about working at the Tutoring Center again.

Strategies

2 Stay positive and confident
6 Use a direct approach—summary of reading
7 Give your opinion if you are asked for it—report of man's and woman's opinions
8 Be sure of your choices—vocabulary and grammar are familiar to the speaker
12 Use the directions as an outline—60 seconds for summary and opinion
14 Use written and spoken contractions—written and spoken used "going to"
15 Use colloquial vocabulary—no slang here
17 Make connections using extension and contradiction—summary of ad and opinion (contradiction)
18 Stop when you have finished answering the question

Task 4

Goal-directed behavior is motivation. If a person's motivated from within then the behavior's *intrinsically* motivated, but if there are external rewards, then the behavior's *extrinsically* motivated. The professor discusses several ways that managers can motivate employees with extrinsic incentives. Flexible schedules, breaks, and lunch times are among the most effective rewards. Another motivator is praise for exceptional work, especially when it's made in a public setting. Personalized rewards are also mentioned. A manager at Bank Boston actually asked his employees to list rewards that would motivate them and one employee included Starbucks coffee. When she finished a big project, the manager wrote her a thank-you note with a gift certificate for Starbucks enclosed. So, the point is that managers have to know their employees well to choose motivational incentives. And money may not be the best one.

Strategies

2 Stay positive and confident
6 Use a direct approach—summary of reading
7 Give your opinion if you are asked for it—factual information with no opinions
8 Be sure of your choices—vocabulary and grammar are familiar to the speaker
10 Support your answers with reasons—reported from lecture
11 Credit the ideas of others—professor cited
12 Use the directions as an outline—60 seconds for concept and case study
14 Use written and spoken contractions—written contractions throughout
15 Use colloquial vocabulary—no slang here
17 Make connections using extension and contradiction—reading concept and lecture example (extension)
18 Stop when you have finished answering the question—strong concluding sentence

Task 5

The woman's problem is that she only has six tickets for graduation but she needs ten so her family can attend. The man recommends two solutions. She could either ask her friends to give her tickets that they don't plan to use or she could put an ad on the college website offering to buy tickets for five dollars each. She's reluctant to impose on her friends, and she's worried that no one will respond to an ad. Well, I think she should call her friends right away. If they don't have extra tickets, then they might know someone who does. She's concerned that her friends might need their tickets, but I agree with the man that a week before graduation, people will probably have their plans made. Besides, if her friends don't have tickets, she can always post the ad online as a last resort. With only one week to go, she needs to solve the problem.

Strategies

2 Stay positive and confident
6 Use a direct approach—summary of problem
7 Give your opinion if you are asked for it—report of suggestions and your opinion
8 Be sure of your choices—vocabulary and grammar are familiar to the speaker
11 Credit the ideas of others—recommendations by man
12 Use the directions as an outline—60 seconds for problem, suggestions, opinion
14 Use written and spoken contractions—written contractions throughout
15 Use colloquial vocabulary—no slang here
18 Stop when you have finished answering the question—conclusion with opinion

Task 6

The lecture explains nanotechnology, which is the investigation of structures that are smaller than 100 nanometers. The scale's so small that a nanometer can be compared to something 100,000 times thinner than a hair. Or, a motor the size of a nanometer would be so small that 6,000 of them would fit on the head of a pin. Recently, the technology's allowed us to observe these small sizes. What's important is that the properties of material change at the smaller scale. For example, gold turns into a liquid at room temperature and copper becomes transparent. So researchers are now trying to control matter to make products with beneficial properties such as faster computers, smaller solar cells, and tiny medical devices. But the benefit comes with some risk.

Strategies

2 Stay positive and confident
6 Use a direct approach—main idea of lecture
7 Give your opinion if you are asked for it—factual information with no opinions
8 Be sure of your choices—vocabulary and grammar are familiar to the speaker
10 Support your answers with reasons—reported from lecture
11 Credit the ideas of others—lecture cited
12 Use the directions as an outline—60 seconds for summary
14 Use written and spoken contractions—written contractions throughout
15 Use colloquial vocabulary—no slang here
18 Stop when you have finished answering the question—strong concluding sentence

CHAPTER 5—STRUCTURE STRATEGIES

PRACTICE 1

1. Hermit crabs ~~carrying~~ *carry* the shells of other species for protection from predators.
2. Based on current predictions, the Earth's population will ^ *be* more than nine billion people by 2040.
3. Whereas planets are formed as a result of natural star formation processes, stars ~~emerging~~ *emerge* from large clouds of gas and dust.
4. ✔ Rice is a labor-intensive plant that grows in flooded fields and thrives in a hot, humid climate.
5. ✔ Koalas, known for sleeping during the day, actively feed on Eucalyptus leaves at night.

PRACTICE 2

1. Because blood from different individuals will have different types of antigens on the surface of the red cells and antibodies in the plasma, a dangerous reaction can ~~to~~ occur between a donor and a recipient in a blood transfusion. *delete "to"*
2. ✔ In the tundra, beneath a rocky layer of topsoil, a layer of permafrost has remained permanently frozen so that the soil can never get soft or warm enough to cultivate plants.
3. An artist who is ~~created~~ *creating* a fresco must have a very clear design prepared in advance because the plaster tends to dry fairly quickly and it is not possible to paint over it with the same ease as it is in an oil painting.

4. The recognized Native American tribes in the United States have ~~form~~ *formed* their own governments on reservation lands.

✔ 5. Some herds have been returning to the same watering holes for decades, encouraging the belief that elephants never forget.

PRACTICE 3

1. Since humans do not inhabit most of the islands in the Galapagos chain, ~~that~~ *delete "that"* the animals are not afraid of people.

2. Most democracies employ a system of checks and balances to ensure that none of the branches of government *become* too powerful.

✔ 3. The Homestead Act provided 160 acres to pioneers who remained on and improved the land in Western territories of the United States.

4. A researcher at the University of Toronto has developed a contact lens *that* can monitor blood sugar levels in diabetics.

✔ 5. Some astronomers claim that more than 90 percent of the mass of the universe is dark matter which can't be seen.

PRACTICE 4

1. Columbus, one of the state's largest cities, ~~are~~ *is* not only the capital of Ohio but also a typical metropolitan area, often used in market research.

2. The cerebrum, which is comprised of both right and left hemispheres, ~~control~~ *controls* the muscles of the opposite side of the body.

3. Influenced by the Phoenician system, the Greek alphabet, which includes 24 letters, ~~were~~ *was* developed in 1,000 B.C.E.

✔ 4. Because some varieties of bacteria produce hydrogen, they are being considered as a potential energy source.

✔ 5. Flutes carved from bone are among the earliest musical instruments found in archaeological sites.

PRACTICE 5

✔ 1. Constellations tend to change over time because every star is constantly moving through space and the pattern disappears.

thinking

2. Repression occurs when an individual avoids ~~to think~~ about painful experiences, and can even temporarily forget about them.

to remain

3. In some optical illusions, an image seems ~~remaining~~ after the exposure to the original image has ended.

smoking

4. Most doctors agree that when patients stop ~~to smoke~~, their lungs repair themselves within seven years.

✔ 5. The Phoenix is a mythical bird that appears to emerge from the ashes of its funeral pyre.

PRACTICE 6

was OR is

1. Cupid, one of the ancient Roman gods, ^ represented as a little child with wings.

divided

2. Our solar system is ~~divide~~ into three categories of planets based on their size and density.

✔ 3. Although a baby is born with 300 bones, several will fuse together by adulthood when only a few more than 200 will normally be accounted for.

are

4. More than forty countries ^ found in Asia, which is the world's largest continent.

✔ 5. The first bank-issued credit cards were offered by the Flatbush National Bank of Brooklyn, New York.

PRACTICE 7

✔ 1. It is essential that antioxidants, which are absorbed primarily from fruit and vegetables, be present to repair cells.

be

2. It is important that renewable energy resources ~~are~~ explored in order to provide alternatives to fossil fuels.

repeal

3. In 1773, after the Americans demanded that the British ~~repealed~~ the tax on tea, their refusal instigated the famous Boston Tea Party.

delete "can"

4. It is vital to the pollination of many crops that researchers ~~can~~ discover why colonies of honeybees are dying.

✔ 5. According to studies, infants prefer that an object have figures and patterns on it rather than solid colors.

PRACTICE 8

learn OR will learn

1. If children are healthy, they ~~are learning~~ to walk at about eighteen months old.

is

2. If the trajectory of a satellite ~~will be~~ off at launch, it will get worse as the flight progresses.

diffuses OR will diffuse

3. If light strikes a rough surface, it ~~diffused~~ because the rays are not parallel to each other.

✔ 4. If the President ignores a bill while Congress is in session, it will become a law after ten days, even without a signature.

✔ 5. If humans are deprived of sleep, the immune system weakens.

PRACTICE 9

That

1. ^ Comets have periodic orbits was confirmed in 1758 by the appearance of a comet when Halley predicted it.

✔ 2. That Michelangelo was highly regarded by his contemporaries is evidenced by the fact that two biographies were published during his lifetime.

delete "That"

3. ~~That~~ the discovery that about 8 percent of the Earth's crust is made of aluminum caused the price to fall from $20 to $1 in the late 1800s.

✔ 4. That seventy-six million American children were born between 1945 and 1964 is significant because the demographic affects marketing trends.

That *delete "that they"*

5. ^ Shakespeare's plays ~~that they~~ were written by another author has not been supported by Elizabethan documents.

PRACTICE 10

is

1. It ^ believed that no two tigers have exactly the same pattern of stripes.

delete "That"

2. ~~That~~ it is proposed that the sun accounts for 99 percent of the matter in the solar system.

✔ 3. It is accepted that maps distort the size and shape of the geographic features because they are flat.

✔ 4. It is well known that the best quality charcoal for sketching pencils is made from vine wood heated in a kiln until only the carbon remains.

It

5. ~~That~~ is thought that the Moon may have had life on it at one time because there is evidence of water at the bottom of the Cabeus Crater.

PRACTICE 11

✔ 1. Found in a wide variety of habitats, some snails have been known to hibernate for three years.

for

2. Hadrian's Wall has survived ~~since~~ 2,000 years, a tribute to the architects and engineers as well as to the Roman armies that built it.

✔ 3. Forest biomes have been evolving since about 420 million years ago.

for

4. Although it influenced other styles of art, Impressionism in its purest form lasted ~~since~~ only about 15 years.

5. New Zealand was the first country to grant women the vote, a right which they have enjoyed ~~for~~ 1893, almost 40 years before women were granted the vote in the United States.

since

PRACTICE 12

delete "the"

1. About 98 percent of Antarctica is covered by ~~the~~ ice, extending over an area of fourteen million square kilometers.

2. Although hundreds of languages are recognized in India, Hindi is considered the primary official language and ~~the~~ English is recognized as a secondary official language.

delete "the"

✔ 3. For thousands of years, wood has been used for both fuel and construction material in areas where trees are plentiful.

✔ 4. Flooding the fields for three days or less is a traditional method for controlling weeds in many parts of the world where rice is cultivated.

delete "the"

5. Evidence of ~~the~~ furniture survives from the Neolithic period, including stone beds, chairs, tables, and dressers.

PRACTICE 13

✔ 1. The Cabinet and the White House staff provide advice to the President.

information

2. Because so many mobile devices are available at relatively low cost, ~~informations~~ from international sources is available instantaneously.

homework
3. College students should calculate two hours of ~~homeworks~~ for every hour of class time that they schedule.

4. From space, astronomers use special telescopes with cameras and other ~~equipments that allows~~ them to study X-ray and gamma ray emissions.
equipment that allows

✔ 5. Works [artistic creations] of music cannot be copied without permission because an artist's work [all of their works] is protected by law.

PRACTICE 14

Carving
1. ~~Carve~~, using either wood or stone, is one of the oldest and most respected crafts worldwide.

✔ 2. Recent studies suggest that sleeping for one hour during the day may not disturb the nightly routine as previously thought.

delete "To"
3. ~~To~~ advertising has changed in response to the increased popularity of Internet sites and mobile devices.

delete "the"
4. According to paleontologists, scientists who study fossils, ~~the~~ classifying dinosaurs can be complicated because they demonstrate a very large range of characteristics.

✔ 5. Although using solar energy does not produce air pollution and water pollution, the materials and chemicals employed in the manufacturing process of the cells do impact the environment indirectly.

PRACTICE 15

isolated
1. Archipelagos, a chain or cluster of islands, tend to form in ~~isolating~~ parts of the ocean.

astonishing
2. After years of watching silent films, in 1927 audiences saw *The Jazz Singer*, an ~~astonished~~ full-length feature with dialogue and singing.

✔ 3. Light pollution from bright, artificial lights can affect ecosystems because it can cause insects and birds to become disoriented.

challenging
4. Some of the most ~~challenged~~ mountains to climb in North America can be found in the Canadian provinces of British Columbia and Alberta.

✔ 5. The origin of the word *Google* is interesting because it was inspired by *googol*, the number 1 with one hundred zeros after it.

PRACTICE 16

✔ 1. Within the wolf pack, the higher the rank, the taller the wolf stands, holding its head, ears, and tail erect to intimidate the others.

✔ 2. Very high mountains, like Pikes Peak, can be seen from as many as 80 miles away, which is about the distance to Limon, Colorado.

as

3. One colony of penguins can number as many ^ 60,000 birds.

Amost all OR Most

4. ~~Almost~~ of the traditional developmental psychologists have focused on childhood.

the

5. Although caffeine is a stimulant, the more coffee the subjects drank in the study, ^ less energy they reported after one hour.

PRACTICE 17

1. Although it is most often associated with the mining of diamonds, South Africa is the world's largest ~~produce~~ of platinum.

producer

aggreement

2. A merger is an ~~agree~~ for two firms to continue in business as a single new company, rather than to remain separately owned and operated.

✔ 3. Because frogs drink and breathe through their skin, they are vulnerable to air and water pollution.

rotation

4. In 1884, a system of 24 standard time zones was adopted to calculate the ~~rotate~~ of the Earth by 15 degrees each hour.

✔ 5. The authorization for most computer security systems is based on a two-step process to verify the user's identity.

PRACTICE 18

heat

1. The ~~hot~~ generated inside the Earth was much greater during the early history of the planet, due in large part to impacts, gravitational pull, and radioactive decay.

✔ 2. Diet, stress, and heredity are a few of the triggers for frequent migraine headaches.

✔ 3. It is currently estimated that our solar system orbits the galactic center once every 250 million years.

4. According to social learning theory, exposure to aggressive role models on television and in films can ~~great~~ influence violent behavior.

greatly

5. The use of mixed media can be more ~~interest~~ *interesting* than that of other more traditional styles of artistic expression.

PRACTICE 19

1. Known as the Painted Lady in some parts of the world, ~~it is between Africa and Europe where~~ the brightly colored Red Admiral butterfly migrates *between Africa and Europe*.
2. ~~To be finding~~ *Found* naturally in many kinds of food, minerals are essential to maintain healthy bones, a strong immune system, and normal blood pressure.
3. ✔ Founded in 1636, Harvard was originally called the New College, but it was not until 1639 when John Harvard bequeathed his library and half of his estate to the school that the college began to provide higher educational programs.
4. Embraced primarily as a reaction against the Age of Reason, ~~artists introduced~~ Romanticism *was introduced* not as a style of art so much as a pursuit of beauty intended to arouse an emotional response.
5. ✔ Running at speeds of up to thirty miles per hour, the Arctic hare can keep up with a car in city traffic.

PRACTICE 20

1. The Bamboo Lemur, a critically endangered species, feeds almost exclusively on bamboo, eating not only the shoots but *also* the pith and leaves.
2. The grieving process may be triggered by a number of situations, including *the death of* a loved one ~~dies~~, a chronic condition that affects the quality of life, or even the loss of a job.
3. ✔ Gold is a highly efficient conductor of electricity, which makes it useful in small amounts in cell phones, global positioning systems, and calculators.
4. ✔ Neither Benjamin Franklin nor Thomas Jefferson was in attendance at the First Continental Congress.
5. At more than 21,000 meters, the highest known mountain in the Solar System is not on Earth but ~~it is~~ *on* Mars.

CHAPTER 6—WRITING STRATEGIES

Example answers for the Writing section provide only one possible response to a question or task. Use them to evaluate your answers. Remember that your answers may be different, but correct.

PRACTICE 1

Goal: 10 minutes, 5 errors = 35 words per minute
Estimates: The following chart will help you to estimate your typing skills.

Evaluation	Minutes	Errors	Words per minute
Excellent	10	5	35
Acceptable	15	5	20
Needs practice	20	5	15

Several free typing tutors are available online.
Try *www.typingweb.com* if you need practice.

PRACTICE 2

1. My ideal job would be working for an automobile company.
2. I think that an excellent teacher is knowledgeable and fun.
3. My favorite book is *The Wizard of Oz*, which I read as a child.
4. The most important invention in the world so far is the telephone.
5. The Internet is a positive influence.
6. In my opinion, class attendance should be required.
7. In my view, playing computer games is a good way to alleviate stress.
8. I prefer a traditional classroom.
9. If I could meet any person, I would like to meet Donald Trump.
10. If I could change one thing about my home town, I would build an airport.

PRACTICE 3

1. Childhood is the best time of life.

Agree	Disagree
No responsibilities	Dependent
Innocent	Ignorant
Future open	Restricted

2. Science is the most important subject in school.

Agree	Disagree
Basis for many fields	All subjects balanced education
Better jobs in technology	Work in area of interest
Knowledge changing	Some schools for arts

3. Computers have improved communication.

Agree	Disagree
Faster	Impersonal
Cheaper	Confusing
More often	Frustrating

4. Playing games is a waste of time.

Agree	Disagree
Play longer than planned	Relax from stress
Have nothing to show for it	Have fun with others
Can be addictive	Learn skills

5. Endangered species should be protected.

Agree	Disagree
Species have right to live	Natural process
Extinction upsets ecosystem	Uncertain results
Plants provide medicines	Interrupted progress

6. Parents are the best advisors for their children.

Agree	Disagree
Love children	Not the same generation
Have more experience	Unfamiliar with options
Understand son or daughter	Place restrictions

7. It is better to read non-fiction books than novels.

Agree	Disagree
Information	Entertainment
Self-improvement	Life lessons
Career	Culture

8. First impressions of people are usually correct.

Agree	Disagree
Appearance	Character
Interaction	Deception
Manners	Introverted personality

9. Celebrities receive too much money for their work.

Agree	Disagree
Exaggerated lifestyle	Charity
Addictions	Role models
Pay per movie or game	Limited time to work

10. Exercise is the best way to manage stressful situations.

Agree	Disagree
Physically relaxes muscles	Some people don't enjoy exercise
Provides social interaction	Meditation just as effective
Gives mind a break	Solving problem provides relief

PRACTICE 4

1. My older brother is the person that I most admire. I feel that way about him because he is the head of our family, a respected member of the community, and a successful professional.
2. If I were to receive a million dollars, I would be amazed. My plan for using the money would include paying my student loan, investing part of the money, and helping my family with the rest.
3. The announcement that a factory will be built on the former site of a park in my hometown is not good news. I am distressed to hear about this development because many people who used the park will not have a place for recreation, the neighborhood near the park will decline in value, and the pollution from the factory will affect the health of the residents.
4. I would rather live in the city than in the country because of the opportunities available in an urban setting. Perhaps the most important advantages are the opportunity to take classes at colleges and universities, the option of using public transportation, and the chance to locate food from my country at restaurants and grocery stores.
5. Although a case could be made for small families, I would rather have a large family. A large family is important because it gives you assistance when you need it, a group to have fun with, and relationships that last a lifetime.
6. When I choose a friend, I look for a number of characteristics, but probably the most important to me is trustworthiness. I value this quality in a friend because you can share confidential information, you know that your friend will tell you the truth, and you won't get into trouble because of your friend's actions.
7. While I could discuss several ways that the Internet has changed the world, three ways seem most significant. The Internet has made research much easier, communication cheaper, and entertainment more accessible.
8. The government has passed legislation to ban smoking in public places. I think that this law is appropriate because of the health hazard for non-smokers, the damage to the public buildings and contents, and the expense to clean up after smokers.
9. Although school is most often associated with education, there are other ways to learn valuable lessons. Three ways that seem worth exploring are the advice of older friends and family, the experience gained by working, and the mistakes that provide helpful insights.

10. Given the opportunity to work for a company or start my own business, I would definitely choose to be a small business owner. The idea of starting my own business is appealing to me because owners can make decisions about the business, set their own work schedules, and earn more money than they could as employees in someone else's company.

PRACTICE 5

Although many people prefer to study in a group, studying by myself is more effective because it saves time, allows me to focus on my weak points, and eliminates scheduling.

1. Study groups often waste time talking about unrelated matters.
2. The group tends to identify areas to study that I don't need to review.
3. It is often difficult to schedule a convenient time to meet.

Although many people prefer to study alone, studying in a group is more effective because it increases my study time, helps me to clarify my notes, and offers me a support system.

1. The commitment to meeting the group keeps me on track.
2. Sharing notes and ideas tends to fill in the gaps from lectures.
3. Moral support and encouragement is helpful when I get stressed.

PRACTICE 6

Attendance should be part of the grading system in university courses. In my opinion, the policy is appropriate because many students will not make mature choices, they profit from a classroom review of the material, and they prepare for other life experiences by respecting an attendance requirement.

Major Point 1
Because many university students are still immature, the requirement encourages them to stay on track.

Supporting Detail: Evidence
One-third of college students will fail during their freshman year, in part because they have not made good decisions.

Major Point 2
Even if some of the lectures are repetitive, it is good to review material that students have read in the book and online.

Supporting Detail: Example
Review is recommended as part of most study plans, as for example, SQ3R (Survey, Question, Read, Review, and Recite).

Major Point 3

Since mandatory attendance in classes is not very different from obligations in the workplace, the policy prepares students for life after they leave the university.

Supporting Detail: Experience

In my experience, most employers do not tolerate employee absences because of expense and inefficiency.

PRACTICE 7

Spoken English:

In selective listening experiments, subjects are presented simultaneously with two spoken messages, but most of the time they are able to recall the message that they are instructed to remember and disregard the other one.

Written English:

In selective listening experiments, subjects are presented simultaneously with two spoken messages; however, most of the time they are able to recall the message that they are instructed to remember and disregard the other one.

Spoken English:

When the Romans invaded Britain, they had to overcome the problem of staging an attack across the English Channel, and they were faced with dissatisfaction within the ranks of their armies.

Written English:

When the Romans invaded Britain, they had to overcome the problem of staging an attack across the English Channel; furthermore, they were faced with dissatisfaction within the ranks of their armies.

OR

When the Romans invaded Britain, they had to overcome the problem of staging an attack across the English Channel; moreover, they were faced with dissatisfaction within the ranks of their armies.

Spoken English:

The Sophists believed that people should take care of themselves, so they advocated a system of education that promoted individual accountability.

Written English:

The Sophists believed that people should take care of themselves; therefore, they advocated a system of education that promoted individual accountability.

Spoken English:

Earth's magnetic field is generated within its molten core. Still several questions remain about the changes in the core's convection.

Written English:

Earth's magnetic field is generated within its molten core; nevertheless, several questions remain about the changes in the core's convection.

Spoken English:
Japan, Taiwan, and Mexico all have early warning systems to provide a short time for the population to take shelter, and many countries are now developing cell phone programs with a warning ring.

Written English:
Japan, Taiwan, and Mexico all have early warning systems to provide a short time for the population to take shelter; furthermore, many countries are now developing cell phone programs with a warning ring.

OR

Japan, Taiwan, and Mexico all have early warning systems to provide a short time for the population to take shelter; moreover, many countries are now developing cell phone programs with a warning ring.

PRACTICE 8

Herbs are plants containing chemicals to treat diseases. The scientific community does not approve of herbal remedies; nevertheless, herbal medicine has traditionally been used in almost every culture. Knowledge of herbal cures has been passed down through the generations in oral histories; furthermore, the ancient cultures of China, Egypt, India, Babylon, and the native people of the Americas preserved their knowledge in early written records. Europeans began publishing their cures in the 1600s.The World Health Organization (WHO) estimates that about 80 percent of the world's population relies on herbal medicine as a primary treatment; moreover, studies show that 75 percent of the plant derivatives in herbal medicines correlate significantly with those in modern treatments.

PRACTICE 9

1. It is remarkable that most individuals have the ability to remember seven discrete chunks of information.

   It is extraordinary that most people have the ability to recall seven separate chunks of information.

2. No universal definition of music exists because the difference between music and noise is identified differently by various cultures.

   No comprehensive definition of music is accepted because the distinction between music and noise is recognized differently by diverse cultures.

3. Vascular plants first appeared during the Silurian period, and by the Devonian period, they had diversified and dispersed into many different land environments.

   Vascular plants first emerged during the Silurian period, and by the Devonian period, they had differentiated and distributed into many varied land environments.

4. One of the oldest surviving examples of casting is a copper frog from 3,200 B.C.E., which was found in Mesopotamia.

   One of the most ancient surviving specimens of casting is a copper frog from 3,200 B.C.E., which was discovered in Mesopotamia.

5. Typically, only one-tenth of an iceberg is visible above the water, which makes it difficult to determine the shape of the part submerged.

   Normally, only one-tenth of an iceberg is observed above the water, which makes it problematic to discern the contour of the part underwater.

6. Polaris, also called the North Star, appears to remain stationary in the northern hemisphere, while other stars and planets revolve around it.

   Polaris, also called the North Star, seems to stay still in the northern hemisphere, while other celestial bodies revolve around it.

7. In some states, an optional preschool year is offered to children of five to six years of age, while in other states, all five-year-olds attend a compulsory year of kindergarten.

   In some states, a voluntary preschool year is offered to children of five to six years of age, while in other states, all five-year-olds attend a mandatory year of kindergarten.

8. Aboriginal Australians were believed to have first arrived on the Australian mainland by boat from the Indonesian Archipelago about 50,000 years before the Europeans.

   Aboriginal Australians were thought to have first landed on the Australian mainland by boat from the Indonesian Archipelago about 50,000 years in advance of the Europeans.

9. Amphibious fish like the mudskipper can breathe out of water for several days and move about on land by manipulating their fins in a skipping motion.

   Amphibious fish like the mudskipper can breathe out of water for a number of days and negotiate on land by maneuvering their fins in a skipping motion.

10. Twins experience a unique language learning situation because, unlike singletons who learn from adults and older siblings, twins receive input from another child at the same level of language acquisition.

    Twins go through an exceptional language learning circumstance because, unlike singletons who learn from adults and older siblings, twins obtain input from another child at the same stage of language learning.

PRACTICE 10

1. Perceptions are based on larger units such as forms, figures, and contexts.

   Negative Report
   The lecturer challenges the theory that perceptions are based on larger units such as forms, figures, and contexts.

2. The similarity in the early developmental stages of all vertebrate embryos suggests a common ancestral form.

   Neutral Report
   The reading passage states that the similarity in the early developmental stages of all vertebrate embryos suggests a common ancestral form.

3. In the Rococo period in Europe, architecture, sculpture, painting, and the decorative arts all exhibited a similar artistic style.

   Certain Report
   The reading passage maintains that in the Rococo period in Europe, architecture, sculpture, painting, and the decorative arts all exhibited a similar artistic style.

4. The problem with personality typologies is that they tend to disregard situational variables.

   Neutral Report
   The lecturer mentions that the problem with personality typologies is that they tend to disregard situational variables.

5. Animal experimentation is not reliable because the underlying physiology of human beings is very different from that of lab animals.

   Certain Report
   The professor argues that animal experimentation is not reliable because the underlying physiology of human beings is very different from that of lab animals.

6. Material aspects of culture are more readily diffused than ideas or abstractions, and even then, it is rare for something to be borrowed intact.

   Doubtful Report
   The reading passage claims that material aspects of culture are more readily diffused than ideas or abstractions, and even then, it is rare for something to be borrowed intact.

7. Craftsmen in colonial America used designs and techniques that were popular in England.

   Certain Report
   The reading passage concludes that craftsmen in colonial America used designs and techniques that were popular in England.

8. The experimental design for many of the studies in parapsychology is flawed.

   Negative Report
   The lecturer contradicts the idea that the experimental design for many of the studies in parapsychology is flawed.

9. About three billion years ago Mars changed from a hospitable planet to a frozen desert.

   Doubtful Report
   The reading passage alleges that about three billion years ago Mars changed from a hospitable planet to a frozen desert.

10. Dancing began as part of the rituals that were practiced by ancient societies.

    Neutral Report
    The reading passage explains that dancing began as part of the rituals that were practiced by ancient societies.

PRACTICE 11

1. It is obvious that everyone should attend college.
   All things considered, everyone should attend college.

2. The university administration should never assign roommates to students.
   The university administration should probably not assign roommates to students.

3. I am sure that paying teachers better salaries would improve their performance.
   I believe that paying teachers better salaries would improve their performance.

✔ 4. If I could study in another country for one year, I would probably choose Canada.

5. Tests always serve a beneficial purpose in education.
   Basically, tests serve a beneficial purpose in education.

✔ 6. I tend to prefer traditional classroom instruction to new technological options.

7. I am certain that cooperation is better than competition in the classroom.
   On the whole, cooperation is better than competition in the classroom.

8. A large university is definitely a better choice than a small college.
   A large university is usually a better choice than a small college.

✔ 9. Teachers probably assign more homework than necessary.

✔ 10. For the most part, learning should be fun.

PRACTICE 12

Conjoined Sentence
Contrast: *however*
Until recently, we assumed that they hunted alone.
Canadian paleontologist Philip Currie has a new hypothesis.

Until recently, we assumed that they hunted alone; however, Canadian paleontologist Philip Currie has a new hypothesis.

Relative Clauses: *that, which*
He thinks that these large dinosaurs hunted in groups.
He calls these groups *dino gangs*.

He thinks that these large dinosaurs hunted in groups, which he calls *dino gangs*.

Conjoined Sentence
Addition: *moreover*
In excavation sites in Canada, many skeletons have been found in close proximity.
Other sites in Argentina also uncovered the remains of groups.

In excavation sites in Canada, many skeletons have been found in close proximity; moreover, other sites in Argentina also uncovered the remains of groups.

Compound Sentence
Addition: *and*
Currie and associates interpret the skeletal remains as proof of social behavior.
They assert that Tyrannosaurs lived and hunted together.

Currie and associates interpret the skeletal remains as proof of social behavior, and they assert that Tyrannosaurs lived and hunted together.

Series: Commas
She proposes three alternative reasons why specimens were found together.
The three reasons include natural disasters.
The topography of the area is a reason.
Famine is a reason.

She proposes three alternative reasons why specimens were found together, including natural disasters, the topography of the area, and famine.

Complex Sentence
Cause/result: *because*
The topography of the area could explain the presence of many skeletons.
Soft mud or quicksand could trap animals.

The topography of the area could explain the presence of many skeletons because soft mud or quicksand could trap animals.

Complex Sentence
Condition: *if*
Philips could provide evidence of groups of footprints that appeared to be those of a group of hunters.
The hypothesis would be more compelling.

If Philips could provide evidence of groups of footprints that appeared to be those of a group of hunters, the hypothesis would be more compelling.

Compound Sentence
Contrast: *but*
Pack hunting is a possibility.
It is not the only interpretation of the evidence.

Pack hunting is a possibility, but it is not the only interpretation of the evidence.

According to the reading passage, Tyrannosaurs were among the most intimidating predators. Until recently, we assumed that they hunted alone; however, Canadian paleontologist Philip Currie has a new hypothesis. He thinks that these large dinosaurs hunted in groups, which he calls *dino gangs*. This idea is not exactly new. In excavation sites in Canada, many skeletons have been found in close proximity; moreover, other sites in Argentina also uncovered the remains of groups. The Korea-Mongolia International Dinosaur Project provided more evidence. Six Tarbosaurus of different ages were found in close proximity in bone beds. Currie and associates interpret the skeletal remains as proof of social behavior, and they assert that Tyrannosaurs lived and hunted together. In contrast, the lecturer disagrees with the theory. She proposes three alternative reasons why specimens were found together, including natural disasters, the topography of the area, and famine. Natural disasters can cause animals to run to a location of safety. For example, a flood could cause the Tyrannosaurs to move to higher ground. The topography of the area could explain the presence of many skeletons because soft mud or quicksand could trap animals. Famine could cause the population to congregate and eventually die in sparse feeding grounds. If Philips could provide evidence of groups of footprints that appeared to be those of group hunters, the hypothesis would be more compelling. Pack hunting is a possibility, but it is not the only interpretation of the evidence.

PRACTICE 13

CD 2, Track 16

Listen to part of a lecture in an anthropology class. The professor is talking about mythology.

Although Native Americans are often familiar with the theories of migration proposed by anthropologists, the traditional creation stories handed down in the oral and written histories of the tribes are still respected by many people. For some, the stories are regarded as part of their religious beliefs, not to be interpreted in a literal sense. For others, the stories are historical accounts with as much veracity as the prehistories published by academics.

It is interesting to note that a large body of Native American mythology across many tribes in the New World recount emergence from underground in the sacred areas where their tribes have historically lived. In the Southwest, the Hopi, the Pueblo tribes, the Apache, and the Navajo all claim to have emerged from the Earth. Furthermore, the stories recount that their ancestors have lived in their traditional homelands forever.

Many Native Americans are skeptical of the migration theories. They believe that they have a connection with the land and the mountains, rivers, and valleys that form both the physical boundaries and the spiritual symbols for their stories. The thought that they may have been wanderers from Asia simply doesn't coincide with their traditions or with the worldview that has been created by the narratives.

Synthesis

Summarize: The reading passage proposes several theories of migration from Asia to the Americas. The Beringia theory contends that nomads migrated across the Bering Straits, following game through the Rocky Mountains, across the Great Plains and into what is now the Southwestern United States, then continued their journey by land into South America. Known as the Clovis people because their culture was first identified in excavations near Clovis, New Mexico, their trail was well marked by arrows and spears with unique flutes cut into the flint. Newer theories posit a water route from Europe or across the Pacific. Moreover, excavations that appear to predate the Clovis sites call into question both of these traditional theories of migration.

Transition: In contrast, the lecturer explains that the oral histories and written accounts by many native peoples reject all of the theories put forward by anthropologists.

Summarize: According to the traditional view of a large number of tribes, they have always lived in the Americas. Many tribes recount stories of their emergence from underground. Furthermore, they believe that they are connected to the land where they live, and many natural landmarks have spiritual significance for them.

Relate: Although anthropologists continue their debate about the route that migrants may have taken to inhabit the Americas, for many Native Americans, the argument does not connect with their traditional creation stories and cannot be incorporated into their worldview.

PRACTICE 14

Independent Essay

Optional introduction: I recognize that some people would prefer to use the investment in space exploration for immediate improvements to life on Earth, but I believe that it is important to continue studying the universe because exploration may provide the best long-term solutions to some of the most serious problems that we face on our planet.

Outline sentence: While I could discuss a number of these problems, the pollution of the environment, the depletion of natural resources, and overpopulation come to mind.

Idea 1: First, it is important to acknowledge that while scientists are exploring space, serious environmental studies of Earth are also being conducted. For example, when scientists travel into space, they also collect samples of air to determine current air quality in the atmosphere and to track changes in our most necessary requirement for life on Earth—the air we breathe. Furthermore, information about climate change can best be examined from space. The movement of bodies of water, the thickness of ice, and the flow of jet streams and weather patterns are all studied during space exploration.

Idea 2: In addition, exploring other worlds could result in the discovery of valuable natural resources. These resources may include many energy sources that are currently being depleted on Earth or even presently unknown resources that could lead to a total revolution in the way that we design our living spaces, vehicles, and technologies. Few would disagree that we are consuming the resources that we have available on Earth. If we do not explore extraterrestrial alternatives, eventually, we will run out of options.

Idea 3: Finally, it is obvious that the population on Earth is continuing to increase. Almost 7 billion people now share the planet, and some argue that the current figures already represent a strain on the number that our planet can support. If we could locate a planet or moon that would support life as we know it, colonization would be an obvious solution to overpopulation. We should not wait for an emergency to explore new places to settle.

Conclusion: Therefore, although funding for the infrastructure and social programs on Earth may seem more attractive and would certainly show a more immediate return on investment, space exploration offers not only the chance to study outer space but also the opportunity to find better solutions to problems on Earth.

PRACTICE 15

Agree or disagree with the following statement: States should not use lotteries to raise money.

1. In the first place, the odds of winning a lottery are very slight.
2. In addition, many people who cannot afford to spend money on the lottery will buy tickets.
3. Finally, gambling can become addictive.

Conclusion: To summarize, states should not use lotteries to raise money because of the odds, the expense for those who purchase the tickets, and the risk of addition to gambling.

Based on the reading passage and the lecture, was Shakespeare the author of the plays attributed to him?

1. According to the reading, the historical record shows Shakespeare's name on the plays and on related documents in reference to the plays.
2. Furthermore, contemporaries of Shakespeare did not doubt the authorship.
3. The lecturer points out that Shakespeare was part owner in the Globe Theater and also a principal actor in the company.

Conclusion: To recapitulate, Shakespeare was most probably the author of the plays because his name appears in historical records, contemporaries attributed the plays to him, and he was involved as an owner and actor at the Globe Theater where the plays were first produced.

Agree or disagree with the following statement: College students should be required to take physical education classes.

1. The fees for physical education classes are the same as those for important academic subjects.
2. The time that is spent in physical education classes could be better spent studying for tests and preparing papers.
3. College students get enough exercise by walking or biking to class or by participating in recreational activities.

Conclusion: In closing, I believe that college students should not be required to take physical education classes because their money and their time could be directed to more important academic pursuits.

What are the characteristics of good leaders?

1. Perhaps the most important characteristic of good leaders is the ability to surround themselves with talented people.
2. Good organizational and management skills are also essential to leadership.
3. Last, but equally important, good leaders must possess the ability to inspire others to action.

Conclusion: In conclusion, a good leader knows how to recruit, manage, and inspire others.

Based on studies reported in the reading passage and the lecture, bilinguals demonstrated certain advantages because of an efficient executive control system in the brain's network.

1. In the first place, bilingual subjects tended to multitask more easily than monolingual subjects, and they made fewer errors while attending to the tasks.
2. Moreover, bilingual subjects demonstrated superior problem-solving skills when presented with both verbal and non-verbal problems.
3. Creativity, as measured by divergent thinking tasks, was also an area in which bilinguals excelled.

Conclusion: To reiterate, bilinguals in the studies excelled at multitasking, solving verbal and non-verbal problems, and divergent thinking.

PRACTICE 16

1. **Map**

I do not agree that everyone should attend college.

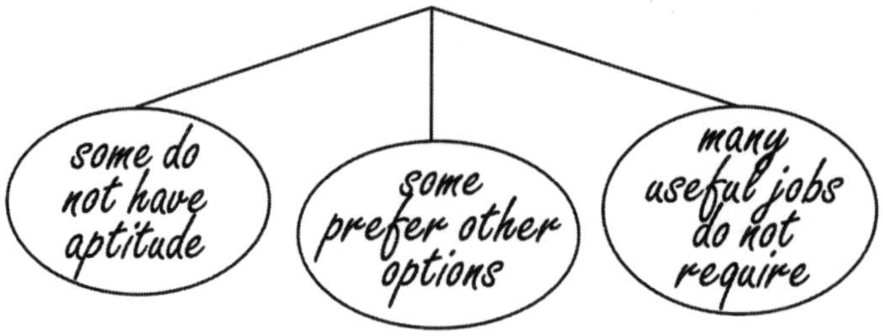

2. **List**

I agree that the university administration should assign roommates to students.

New students do not know other students well enough to choose a roommate.
Computer information can match students with similar backgrounds and interests.
The university experience includes interacting with other college students.

3. **Outline**

I. I disagree that paying teachers better salaries would improve their performance.
   A. Most teachers are trying to do their best already.
   B. Teachers have not entered the profession for the money.
   C. Money is not the only incentive for improved performance.

4. **Map**

If I could study in another country for one year, I would choose to go to Canada.

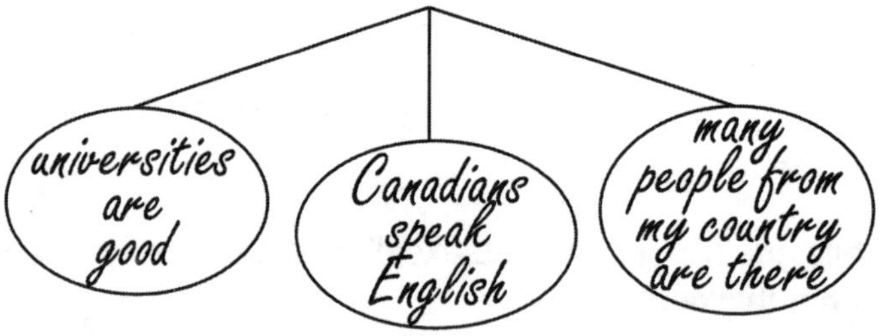

5. **List**

Although standardized tests do not always provide an accurate indication of a student's abilities, I think that they serve a beneficial purpose in education.

They allow a large number of students to be compared with each other.
They give educators one indicator of a student's aptitude or proficiency.
They are scored objectively.

6. **Notes**

Although new technologies have many advantages, I prefer to study in a traditional classroom.

Meeting TC regular schedule

Interacting face-to-face w/ teacher + other students = more enjoyable

TC ≠ learn technology w/ non-technical subject

7. **List**

I agree that cooperation is better than competition in the classroom.

Cooperation in school prepares people for cooperation in the workplace.

Many people learn better by talking and interacting with others.

Cooperation improves relationships.

8. **Outline**

I. I would prefer a small college to a large university.
   - A. The professors at a small college are better because they are more interested in teaching than research.
   - B. The classes have fewer students in them, which allows more interaction with the professor.
   - C. Socializing and studying with a smaller number of people encourages friendships.

9. **Map**

I agree that teachers assign more homework than necessary.

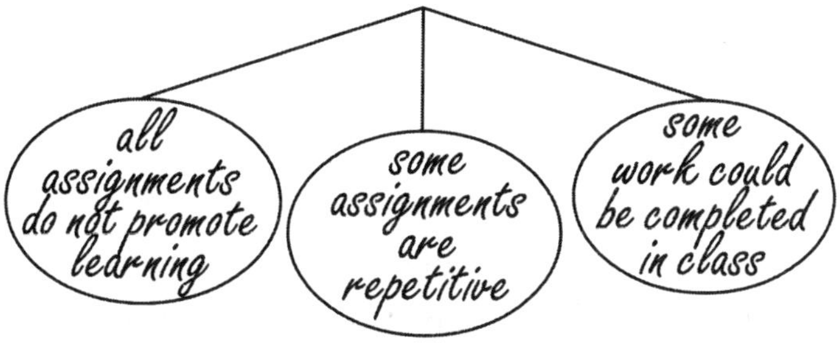

10. **Notes**

Although many people feel that learning should be fun, I do not believe that it is important to have fun in the classroom.

T ≠ required entertain S

S everything in life not fun

activity fun for some Ss not fun for others

PRACTICE 17

Synthesis

The reading passage described a classic psychology experiment. Philip Zimbardo and colleagues at Stanford University designed a study to determine how participants would respond to a simulation of a prison environment. Volunteers for the study are white middle-class males who had no criminal record and no psychological issues. The question was whether good people would retain their normal reactions when placed in a prison environment.

The simulated prison consists of three cells, six by nine feet, with three cots in each cell to accommodate three prisoners. Other rooms across from the cells were designated for the prison guards and the warden. The volunteers were randomly selected to participate as either a prisoner or a guard. Prisoners were arrested, searched, fingerprinted, instructed to pose for mug shots, dressed in typical prison smocks with numbers on them and a stocking cap. They were also chained around their ankles. Guards were dressed in khaki uniforms and were provided with sunglasses to eliminate eye contact with the prisoners. They carry wooden batons, which identifies their authority, but were not to be used to punish prisoners. Guards were to identify each prisoner by his number, not by name. Prisoners are confined to their cells for the duration of the study, whereas guards were assigned to eight-hour shifts, and were free to return home between shifts. Researchers observe and record behavior using cameras and microphones concealed in the experimental areas.

The lecturer reported that guards were free to interact with the prisoners as they deemed appropriate, but many of them become aggressive or even abusive toward the prisoners. After a riot on the first day to protest the stocking caps that they were forced to wear, most of the prisoners became withdrawn and passive. Although the experiment was originally designed for 14 days, five prisoner participants left the experiment early, one research assistant objected to the abusive situation, and the entire experiment is discontinued after only 6 days because the remaining prisoners were beginning to demonstrate extreme stress and anxiety. In conclusion, the Stanford Prison Experiment demonstrated the very powerful influence of situations in human behavior. Only a few people are able to behave according to their usual moral dictates.

are	were
consists	consisted
carry	carried
identifies	identified
are	were
observe	observed
record	recorded
become	became
is discontinued	was discontinued
are	were

PRACTICE 18

Although some students prefer a traditional classroom, I like to take classes online. These nontraditional classes offer the following advantages [1] greater flexibility [2] increased communication, and lower cost.

Perhaps the most important attraction for me personally is that an online class schedule can be adjusted to fit my routine. Since I work full time, traditional courses are often offered when I am obligated to be on the job [3] In addition, I am married with two children, and my family deserves my attention and attendance at school events. [4] time that I would spend driving to campus and finding a parking space can be used for work and family activities. It is not uncommon for me to study at night after I have finished work and the children are in bed.

[5] Another advantage for online instruction is the opportunity to communicate more effectively with the teacher and other students. I tend to be rather shy about speaking in class, especially in a large lecture setting.

Online classes offer me the opportunity to ask questions by e-mail. I also notice that I interact more with my classmates because chat rooms and threads online support communication. In a traditional class, only a few students tend to talk, but online more of us are asking questions and responding to each other [6]

Tuition for an online class varies with the school and degree program in much the same way that it does for a traditional class, but in my experience, the cost of books and lab materials is lower for online classes. [7] many programs do not require textbooks, which tend to be very expensive. On the contrary, the materials for classes and labs are often available as part of the fees for online instruction and can be downloaded along with lectures [8] handouts, additional notes, and other resources.

[9] I do not claim that online classes should replace traditional classes for everyone. In fact, a very good case could be made for them at certain stages of life. What I maintain is that for me, at this point in my life, online classes offer the better opportunity for me to study while I am meeting my work obligations and my family responsibilities [10]

1 advantages:
2 flexibility,
3 job.
4 Time
5 Indent paragraph
6 other.
7 Many
8 lectures,
9 Indent paragraph
10 responsibilities.

PRACTICE 19

Independent Essay 1
Level: 5
Rater's Comments:
This essay demonstrates the writer's ability to organize an essay. The first sentence is direct and responds to the topic. The thesis statement is supported by a detailed example. The meaning is easy to understand. Vocabulary is appropriate, sentence structure is varied, and errors in grammar are minor. A good conclusion is developed in the last paragraph. Although the essay is longer than required, it is logically connected and well written.

Independent Essay 2
Level: 4
Rater's Comments:
The writer responds to the topic and the task. The essay is well organized and the sentences are logically connected. The sentence structure is not as varied as a higher-level essay. Errors in grammar and word choice are found throughout the essay; however, they do not interfere with the reader's ability to understand the meaning. The last sentence provides a good concluding statement. The essay is the minimum length required.

Integrated Essay 1
Level: 5
Rater's Comments:
A wide range of vocabulary and varied structures are used in this well-organized and well-developed essay. Sentences are logically connected and supported by examples and details from the reading passage and lecture. The content is accurate. Minor errors do not interfere with the reader's comprehension. The length of the essay is appropriate.

Integrated Essay 2
Level: 3
Rater's Comments:
The writer responds to the topic and the task in a partially organized and developed essay. Although the content appears to be accurate, it is somewhat incomplete and unclear. Sentence structure is limited with frequent grammatical errors and a limited range of vocabulary. The meaning requires some interpretation by the reader. The length of the essay is slightly shorter than the minimum length required.

PRACTICE 20

Close your eyes. Take a few minutes to visualize an image of your goal. Why are you taking the TOEFL? What will happen for you after you succeed on the test? See yourself achieving the goal that motivates you to take the TOEFL. Enjoy the moment. Practice visualizing this image from time to time as you prepare for the TOEFL. Learn to bring up the image quickly to encourage yourself. Then go back to work. Visualize this image before you begin the last section of the TOEFL. Remember, we become what we believe ourselves to be.

Best wishes for success on the TOEFL and after the TOEFL!
Dr. Pamela Sharpe

Congratulations!

On the following page, you will find a certificate of participation and completion, signed by the author.

Fill in your name.

Completing the strategies in this book is an important step toward achieving your goal.

Certificate of Participation

Awarded to

____

For the successful completion of

Outsmart the TOEFL

BARRON'S

Pamela J Sharpe

Dr. Pamela J. Sharpe

CD1

Track 1: Introduction

Chapter 3—Strategies for the Listening Section

Track 2: Listening Strategy 1 Practice (page 67)

Track 3: Listening Strategy 2 Practice (page 69)

Track 4: Listening Strategy 3 Practice (page 71)

Track 5: Listening Strategy 4 Practice (page 73)

Track 6: Listening Strategy 5 Practice (page 75)

Track 7: Listening Strategy 6 Practice (page 77)

Track 8: Listening Strategy 7 Practice (page 80)

Track 9: Listening Strategy 8 Practice (page 83)

Track 10: Listening Strategy 9 Practice (page 85)

Track 11: Listening Strategy 10 Practice (page 87)

Track 12: Listening Strategy 11 Practice (page 89)

Track 13: Listening Strategy 12 Practice (page 91)

Track 14: Listening Strategy 13 Example Script and Practice (pages 92 and 93)

Track 15: Listening Strategy 14 Example Script and Practice (pages 94 and 95)

Track 16: Listening Strategy 15 Example Scripts and Practice (pages 96–99)

Track 17: Listening Strategy 16 Practice (page 101)

Track 18: Listening Strategy 17 Practice (page 103)

Track 19: Listening Strategy 18 Practice (page 105)

Track 20: Listening Strategy 19 Practice (page 107)

Track 21: Listening Strategy 20 Practice (page 110)

CD2

Chapter 4—Strategies for the Speaking Section

Track 1: Speaking Strategy 1 Practice (page 115)

Track 2: Speaking Strategy 2 Practice (page 117)

Track 3: Speaking Strategy 3 Example Script and Practice (pages 118 and 119)

Track 4: Speaking Strategy 4 Practice (page 121)

Track 5: Speaking Strategy 5 Practice (page 124)

Track 6: Speaking Strategy 6 Practice (page 127)

Track 7: Speaking Strategy 8 Practice (page 131)

Track 8: Speaking Strategy 9 Practice (page 133)

Track 9: Speaking Strategy 12 Practice (pages 144–146)

Track 10: Speaking Strategy 13 Practice (page 151)

Track 11: Speaking Strategy 14 Examples and Practice (pages 154–156)

Track 12: Speaking Strategy 16 Practice (page 161)

Track 13: Speaking Strategy 17 Practice (pages 164–165)

Track 14: Speaking Strategy 19 Practice (page 170)

Track 15: Speaking Strategy 20 Examples and Practice (pages 171–172)

Chapter 6—Strategies for the Writing Section

Track 16: Writing Strategy 13 Practice (page 266)

Track 17: Concluding Remarks